Preface

Many books have been written on the subject of specification writing, including some useful books of reference. However, having taught the subject for many years, the author feels there is a need for a book which can be used as a guide for those writing a specification for the first time, either alone or with some guidance, or for the more experienced to make comparisons in method.

The object of the book is threefold. Firstly to relate the specification to other communicative documents used in the building process. Secondly, to describe types of specification, their respective parts, and their relationship to each other. Thirdly, to direct the reader, either through the use of direct comment, or introductions to trade sections, charts, etc., to other areas of study enabling basic information to be expanded.

The two example specifications are based upon drawings which, with imagination and assistance of alternative material in the text, can be used as skeletal structures upon which exercise can be based.

This book should be regarded as a primer and not as a treatise on the subject of specification writing.

John J. Scott

Contents

Specification Writing (Scott) – Errata

Page 18. Drawing OFB 3 – Section A-A windows to read 5 and 6
Page 49. Drawing TK2 – SP/WP in garage to be transposed (no TG to SP)
Page 53. Drawing TK4 – window numbers to read W9 and W14
Pages 53/55. Drawing TK 4/5 – foundation widths to read 600
Page 59. Drawing TK7 – should read North West Elevation
Page 61. Drawing TK8 – should read South West Elevation

Introduction

The Oxford Dictionary defines specification as, "An article or particular specified, or a detailed statement of particulars, especially workmanship and materials, to be undertaken or supplied by an Architect, builder, manufacturer, etc."

In simplistic terms, the object of a specification is to communicate to someone how something is to be done, so that the specifier's intentions are clearly understood without doubt or ambiguity, and there will be no confusion in the mind of the person who has to perform the specified works.

The term Architect is used throughout this book, but this could be any other person concerned with specification writing. The Joint Contracts Tribunal Standard Form of Building Contract 1980 is referred to throughout. All place names, firm names, products, etc., used in the text are fictitious. Other names given for purposes of reference or information are correct at the time of publication.

All building specification writing starts with the assumption that the writer has a good knowledge of basic building methods and construction, services and materials, and site procedures.

The drawings on pages 13 to 23 and pages 47 to 61 have been produced to a percentage reduction and should not be treated as scale drawings.

1 The contractual and legal implications of specification documents

Standard specifications

Standard specifications are often used in practice, but they may not always be appropriate for all jobs. They are satisfactory in relation to elemental drawings, system-type buildings, and repetitive works. Use of manufacturers' specifications, either directly or by implication, is in order, provided that the contractor and others concerned with the works are aware of the specifier's intentions in relation to their own contractual responsibilities.

The adaptation of existing specifications, the 'cut and paste' technique, can give rise to difficulties and embarrassment on the site, which can be avoided if each job has its own properly written specification. This is not difficult if a logical and orderly approach is adopted.

Brief specifications

The term 'brief specification' is still widely used, but it is really a self contradiction. A brief specification aims to describe the work to be done in simple terms, and should never be used as a basis of contract, or as a working specification on the site. It can be used to help keep the cost planning and cost control factors in step with the initial building design as the project develops.

Schedules of work

Schedules of work are really forms of brief specification; they are less common than hitherto, but are still used on direct-labour works, and where the quality of the ensuing works is of a known and acceptable standard. The description of the works is usually, but not necessarily, of the 'spot item' type of specification described later, but written as briefly as possible. This type of schedule may be used with an agreed schedule of rates where direct labour is not employed, and may also include Prime Cost Sums for fittings and other items, to ensure some control over the standard of work and the cost of such items.

The legal position

The specification document describes the quality of workmanship and the materials to be used in the works. It also provides by means of preliminary items, information and directions to the contractor, who is to estimate for and possibly carry out the works, directly affecting his contractual position and responsibilities. These items are described below. Where the JCT Form of Building Contract (Without Quantities) 1980 is used, the Form of Contract, the Specification, and the contract drawings together form the basis of the contractual agreement between the Client and the Contractor, or the Employer and the Employee, as they are called in the contract, upon which the contractor's tender is based, which if accepted by the Client, becomes the Contract Sum as written in the contract.

In the event of litigation or arbitration arising from disputes in the course of the works, these basic documents are used to determine the contractual intentions of the parties at the signing of the contract. The drawings, Specification, Bills of Quantity, etc., produced by the Architect and other members of the professional team concerned with the work on behalf of the Client therefore have profound legal significance, and failure to safeguard adequately the client's interests in this regard, through inadequate documents or drawings, could lead to actions for negligence against those concerned with their preparation.

Where the JCT Standard Form of Building Contract (With Quantities) 1980 is used the position is different.

The Form of Contract, the Bill of Quantities, and the contract drawings together form the basis of the contract. The Specification is not directly a contract document, it is by implication a supporting document, for if the Bill of Quantities has been prepared properly, it should have been taken off, worked up, abstracted and billed, from the drawings and the Specification read together, with which the Quantity Surveyor should have been provided. There are many reasons why actual practice departs from this ideal, which can cause considerable professional difficulty if things go wrong during the course of the works.

For the Architect the production of the Specification is included in the Preliminary and Basic Services as described in *Architect's Appointment*, published by the Royal Institute of British Architects, although the new Clause 1.16, Work Stages F and G, 'Production Information and Bills of Quantities' is not defined as clearly as in previous documents. The "Small Works" edition S.W.19 is unambiguous.

2 Specification format

The choice of method depends upon the job in hand, its size, and the nature of the works to be specified. There are two ways of writing a specification, a 'spot item' and a 'trade' specification.

Preliminaries

Both types of specifications begin with a separate section called the 'Preliminaries'. This section deals with items which may be called 'contractual risk', 'directives', 'instructions', 'references', 'information', etc. When the work is tendered for, some of these items will attract a sum of money to be entered against them by the Contractor's estimator. Thus, 'Preliminaries' have a direct cost percentage relationship to the total cost of the works, which will vary according to the job in hand. Here are some of the typical items found in the Preliminaries.

Contractual risk item

"The Contractor is to provide all temporary lighting required in the progress of the works, install and maintain same, pay all charges in connection therewith, remove same and make good on completion of the works."

Such an item would attract a sum of money which would be the estimator's assessment for providing this service.

Direction item

"The conditions of the contract shall be those of the JCT Standard Form of Building Contract (Without Quantities) 1980 current edition, and the Contractor is referred to the clauses therein for the purposes of tendering."

Instruction item

"Include the Provisional Sum of £2000 (Two Thousand Pounds) as a Contingency Sum to be used in whole or in part as directed by the Architect, the whole or any part not so used to be deducted from the final account on completion of the contract." (See full explanation of Contingency Sum, page 4.)

Reference item

"*The term BS shall mean the current British Standard Specification of the British Standards Institution, 2 Park Street, London W1A 9BS, 'current edition'. The term BSCP shall mean the 'current' British Standard Code of Practice of the British Standards Institution.*"

Information item

"*Access to the site is from Blackhorse Lane, an unadopted, unmade, private road on the East side of the A362 approaching the village of Little Bagley from Tonbridge.*"

Preliminaries vary from job to job, although several of the items will be repeated on all jobs, and will become familiar with use, e.g. 'Water for Works', 'Attendance', 'Contractor's responsibility'.

In the sample specification at the end of this chapter, a set of Preliminaries is included that would be suitable for a small job, and they will vary according to the particular information that the specifier wishes to convey to the Contractor for the job in hand. You should obtain copies of specifications for various sizes and types of jobs and make comparisons.

Three very important items occur frequently in all specifications, the Contingency Sum, Provisional Sums, and Prime Cost or P.C. Sums.

The Contingency Sum

This is always found in the Preliminaries. It is a sum of money which the specified directs the Contractor to include in his tender for the works, to provide funds for contingency items of work which could not reasonably have been foreseen at the time of writing the specification. This sum of money should not be regarded as a 'sinking fund' for errors and omissions in the contract documents; it is used solely at the discretion of the Architect as he so directs. If it is not used, it is deducted, either in whole or in part, from the final account as a saving on completion of the contract.

Works under the heading of 'Contingencies' are dealt with as variations to the contract (see JCT Standard Form of Building Contract 1980, Clause 13). The Contingency Sum may be 5–10 per cent of the preliminary estimate of the total cost of the works, but will vary according to the nature of the work to be done, e.g. in conversions and similar works, where contingencies are more likely to arise than in new works, the Contingency Sum may be relatively high in relation to the total cost of the works. If the Contingency Sum allowed in the contract is insufficient to cover the contingencies, the contractor may claim an extra over the contract sum as provided for in Clause 13.

Provisional Sums

These sums of money, large or small, are in the description of the works not in the Preliminaries, and should not be confused with the Contingency Sum, which although provisional, is used for a different purpose. A Provisional Sum is a sum of money included to provide for items of work which could not be described when the specification was written. These sums of money may be determined after careful consideration and assessment of the work to be done, plus a safety margin, based upon the experience of other jobs, or simply 'guesstimates', which if badly underestimated, can be an extra over the contract sum, and embarrassing, with a temptation to meet the 'short fall' out of the the Contingency Sum, which is wrong. Here are some examples of Provisional Sums:

> *Works to existing staircase*　　　　*Include the Provisional Sum of £100 (One Hundred Pounds) for work to existing staircase*

This sum is assessed against the actual work done to the staircase, measured as provided for in the contract, when the final account is rendered by the Contractor, and is a direct addition or omission item. It is best to measure and agree the cost of the work immediately it is done to avoid possible later disputes, the work being measured under Clause 13 as previously described.

A large item:

> *Works to existing tunnel*　　　　*Include the Provisional Sum of £5000 (Five Thousand Pounds) for work in sealing off existing tunnel (see Engineer's Drawings No. X1)*

It is unlikely that an item of this size would appear in works without a Bill of Quantities, and it would be transferred from the specification to the Bill of Quantities by the Quantity Surveyor and re-measured on site.

Prime Cost Sums

These are called P.C. Sums, and are the first prime cost of an item before the addition or subtraction of any discount, except for those discounts as provided for in the JCT Standard Form of Building Contract. Prime Cost Sums are used to include for quite small items such as ironmongery, sanitary fittings, wallpaper etc., in order to place a cost limit upon them for the purposes of tendering.

They may also be used to provide for large installations such as heating and electrical works, etc. costing several thousands of pounds, for which the Architect wishes to nominate a specialist sub-contractor.

Prime Cost Sums differ from Provisional Sums as they attract a separate profit item, and in some cases what is called 'attendance', see below.

Here are some examples of Prime Cost Sums:

Electrical installation   *Include the P.C. Sum of £1200 (One Thousand Two Hundred Pounds) for the electrical installation to be carried out by Black & Green Ltd. 4, The Pavement, London, SW4.*
Allow for profit
Allow for attendance

A P.C. Sum such as this is based upon estimates obtained from specialist sub-contractors, and includes a $2\frac{1}{2}$ per cent discount for cash to the Main Contractor as provided for in the contract. The Main Contractor inserts against, and in addition to the P.C. Sum stated, his own assessment of profit and attendance percentages pro rata.

If the Architect does not want to name the firm to carry out the work until later he can say . . .

Allow the P.C. Sum (as before) . . . . to be carried out by a firm to be nominated by the Architect.
Allow for profit
Allow for attendance

This is from where the term 'Nominated Sub-Contractor' is derived. Prime Cost Sums include for VAT where appropriate.

There follows a Clause where the Architect wishes to limit the client's range of wallpaper selection for tendering purposes.

Wallpaper   *Include the P.C. sum of £6.00 (six pounds) per roll, including VAT for wallpaper to be selected by the Architect.*

Such an item attracts no 'profit' or 'attendance' items in addition, as the Main Contractor does the work of preparation and hanging of the paper, which is described separately as builder's work in the specification.

Here a word of warning is appropriate. Do not forget to enter additional clauses for builder's work which may arise in the inclusion of P.C. sums. The 'attendance' where appropriate is limited to the use of site facilities, etc., provision of minimal labour in unloading, stacking and storage of materials until required in the works, and the return of crates, containers, etc. carriage paid. Such items as cutting away, forming chases, bases, building in clips, fixings, lifting and replacing floorboards, etc., are builder's work items. This can be covered by including an omnibus clause in the Preliminaries, but if in doubt insert a clause to cover it. Duplication is always better than omission.

Phrases such as 'supply and fix' are unnecessary as the Contract requires the Contractor to supply everything necessary for the proper completion of the works as the basis of 'lump sum' contracting, i.e. 'A whole work for a given sum of money' in the absence of variations and extra works.

Phrases such as 'to be supplied by' should be avoided, as they may be construed as 'fix only' items by the estimator. It is better to use "to be obtained from" which is unambiguous.

Estimators cannot price accurately from guesswork, and phrases such as 'or other approved' should be avoided. Quantity Surveyors in producing Bills of Quantity also require facts, and the drawings and specification are the means of providing these.

The Preliminaries sample which follows is kept to a minimum, and merely serves as a vehicle for explanation of the kinds of clauses which are used. Items such as 'Shoring', 'Lighting', 'Demolition', etc., with which you will come into contact, must be dealt with as and when they arise.

There is a tendency sometimes to over-write the Preliminaries, and to re-state in different terms things which are already in the contract. This is not recommended, as the contract is written by lawyers in legal language, and the interpretation of the conditions of the contract, when they are disputed, is a matter for the courts to decide. Other clauses written into the specification may give rise to ambiguity and detract from the true meaning of the contract clauses.

Over-written Preliminaries in an attempt to 'screw the Contractor down' to use a time-worn phrase, leads to inflated tender prices. There must be mutual 'give and take', and a measure of trust on both sides. Contractors are in business to make money, they have to ensure that successful tenders absorb the costs of unsuccessful ones, and that overall they make a profit. If the documentation prepared for them to work from is slipshod, they will take advantage of this and charge extras where possible. If the documentation is reasonable, they will want to get on and complete the work promptly in order to use their resources to advantage elsewhere.

Most Contractors, in common with business people as a whole, want to get a continuity of work, and if they can build up a good relationship with Architects and others for whom they work, this is a much better objective to achieve than arguments over what was or what was not intended to be construed from poor drawings, specifications, and other documents.

Do not be afraid, or too proud, to seek advice from or listen to Clerks of Works, Agents, Foremen, Tradesmen, on the site or in the workshops off the site. They are skilled people who deal daily with the practical side of building, and can often suggest ways of doing things which will achieve the results you require satisfactorily, and often more cheaply, than what you have shown on the drawings or described in the specification. This applies particularly to joinery work, which, without a good knowledge of timber, machining techniques, available timber sizes etc., can lead to details which are wasteful of timber, difficult to machine, and expensive.

Conversely, if you see things on site which are not in accordance with the contract documents, raise the matter tactfully with the Clerk of Works, if one is on site, or with the person in charge, but not directly with Tradesmen or Trades Foremen, there is a site protocol to be observed.

3 Preparing to write the specification

Whether the specification is of the 'spot item' or 'trade' type, both will require some preparatory work including a site check. If a preliminary survey of the site or building has been made prior to the preparation of the drawings, useful information can be taken from this for the specification draft. In most cases however, a site visit has to be made, which may be a long way from the office, and may also be in unoccupied premises. Here is a list of basic items to take:

1 a camera;
2 a battery tape recorder (not essential but can be useful for general appraisal of the site and its environs);
3 a hand torch which can be hung up or placed on a flat surface, and a spare bulb;
4 a large screwdriver, and if possible a set of manhole keys. A bradawl for testing suspicious surfaces for wood-rot, etc.;
5 an A4 clip-board with ruled paper, two columns 35 mm wide on the left-hand side;
6 a folding rod or extending rule and a 50-metre steel tape;
7 a sharp knife;
8 two pencils (ball-point and felt-tipped pens can smudge);
9 a pocket compass to obtain correct orientation of the site and building;
10 a plumb line;
11 an Ordnance Survey map of the area with the entire site and environs shown;
12 a sponge bag with cleaning materials.

In conversion works a large amount of the draft specification will have to be written on the site, and in new works site clearance and demolition may be necessary, in both cases a number of items will affect the Preliminaries.

A check list can be compiled in the office and can form a permanent record for the job file. Some of the items included are as follows:

Specification site check list

Job: Old Forge Building, Manor Farm, Frettingham, Norfolk
    Alterations and Additions
Date: April 1983   *Client*.................

Access		The access to the site is from Sandy Lane on the west boundary of the site, a temporary cross-over from the main road into the lane is necessary. Adequate protection for existing trees to be provided at the site entrance, 1 Chestnut 3 m girth, 1 Beech 2 m girth
The lane is an unmade surface (compacted earth and gravel with an uneven surface). Some protection and support for wheel loads will be necessary.
Storage of materials, etc.		No materials can be tipped or stacked in the lane or main road due to restricted width, there is ample space within the site area.
Hoardings		No site enclosure is necessary.
Site location		The site is on the B1150 road 8 miles west of Aylsham and 7 miles east of Norwich.
Services	*Water:*	A 75 mm main runs along the main road and there is a stop cock at the east corner of Sandy Lane.
	Gas:	There is a gas supply to the farmhouse, and this is recorded as being 40 years old.
	Electricity:	A three-phase supply to a transformer exists 110 m west of the site serving a small housing estate. There is an overhead service terminating at the farmhouse with a temporary 230/250 V supply to the forge building.
	Telephone:	An existing service is connected to the farmhouse.
	Drainage:	A 225 mm soil sewer exists in the main road. Surface water discharges into a watercourse on the east boundary via soakaways.
Subsoil		Local authority records indicate loamy clay to an average depth of 6 m, but some made-up ground exists in the area which is unrecorded.
Site environs		The general area of the site and its environs is typically rural, the main road is narrow with no footways, and is a busy road carrying heavy goods vehicles and light traffic from Norwich and Aylsham. Access to the site is restricted both in width and for turning of heavy vehicles, sight lines from the lane to the main road are poor.
Within the site curtilage the area around the farmhouse entrance and the forge building is irregular with some patches of hard standing and some hogging, patches of rough overgrown grass and weeds.
There is a brick built store and toilet with a wash basin and gas water heater adjacent to the East boundary.
There is no public transport to the site, the nearest point being Buxton some two miles distant.

These are typical notes and the list could be very long on some jobs. We can now use some of this information to write the specification.

Vocabulary will come with practice, and specification writers have their own idiosyncrasies, the essential thing is to be brief, but concise and clear in what you wish to say, avoiding trying to describe exactly how things should be done. The Tradesmen, being trained and skilled persons, will know how to perform their respective tasks. They will want to know what is to be done, where it is to be constructed or placed and the standard of workmanship and the materials to be used, plus any additional items such as the provision of fixings for items to be later installed (see timber roof example on page 38).

You will see that there is no mention of bevels, scarf joints, birdsmouths, etc, the Carpenter will execute such labours as necessary. You have a Clause in the specification to cover the workmanship under 'Acceptable building practice' in the Preliminaries.

The following sample is a small conversion and extension job using the 'spot item' method of specification writing. You should carefully check through the clauses against the check list, and relate them to the drawings provided.

With the aid of the 'trade' specification clauses and the charts in the Appendix, specify some alternative forms of construction, different plumbing materials, partitions, roofing, external walls, etc., as an exercise. Some of the abbreviations listed in the Appendix are used in the draft, but the final specification would be written in full.

4 The 'spot item' method of specification writing

Introduction

This method is used where a full trade specification would be too unwieldy. Most jobs of conversion and alteration use this specification method without quantities. It is usual to specify the works under two main headings:

1 *The internal works*, which can be done floor by floor, room by room, whatever seems more appropriate, but following a sequence which approximates to the order in which the contractor may carry out the works.

2 *The external works*, which would include, among other things:
Excavation and site works
Drainage
Paving and landscaping
Fencing
External plumbing
Roof works
Brickwork and renderings, weatherboarding, etc.
Decoration

To avoid repetition of description after the Preliminaries, there is a Preamble to Trades section describing the workmanship and materials to be used, and in the works description following you will see how useful this is, for example:

'*Build 225 mm walls in common bricks in mortar mix B a.b.d.*'

If you refer to the Preambles you will see that the bricks, bond, and mortar are fully described, and whether the work is above or below ground level.

Preliminaries are written in the usual way using the site notes and other information you have to hand. The Preliminaries example given should not be used verbatim, and you should make several attempts to write the Preliminaries for a small job in the office until the 'feel' of the work seems to be right. Get a more experienced colleague to check it for you, as you may have missed the obvious.

The 'spot item' specification when completed will comprise:

1 Preliminaries;
2 the preambles of workmanship and materials;
3 the Works, i.e. the description of the actual work to be done.

If the size of the job warrants it, you may include schedules of doors and windows, ironmongery, finishes, etc. (see examples in the Appendix). On small jobs these would be included in the works description as in the following example.

With all specification writing, start with a carefully prepared check list of all items to be specified, in the approximate order in which they would be carried out. This will help you to visualize the total works, and as far as possible avoid errors and omissions which will occur if attempts to specify directly from the drawings are made. Be critical of your description, remember to be precise, avoiding such phrases as 'should be . . . ', etc.

Old Forge Building, Manor Farm, Frettingham, Norfolk

Notes on existing building

The existing building is about eighty years old, and was used as a forge and storage building for the farm. The forge being disused for many years, with only the recess and flue remaining.

Originally the building was constructed of 225 mm brickwork in soft red bricks and lime mortar, but some five years ago the external walls were rendered in roughcast finish, and at the same time the roof was re-tiled and is in sound condition.

The internal and external walls have a slate damp proof course.

The internal ground floor area is covered with very dense, dry, granite setts.

The upper floor is reached by an external timber staircase, but it is not intended to convert the upper floor. The existing suspended first floor is constructed of heavy sawn joists, open soffited, and covered with heavy section sawn boarding.

There are no windows on the ground floor, and the double doors are of heavy framed, ledged and braced construction.

The existing drain from the farmhouse to the sewer is only a soil drain, the surface water discharges into soakaways and then into a watercourse on the East boundary.

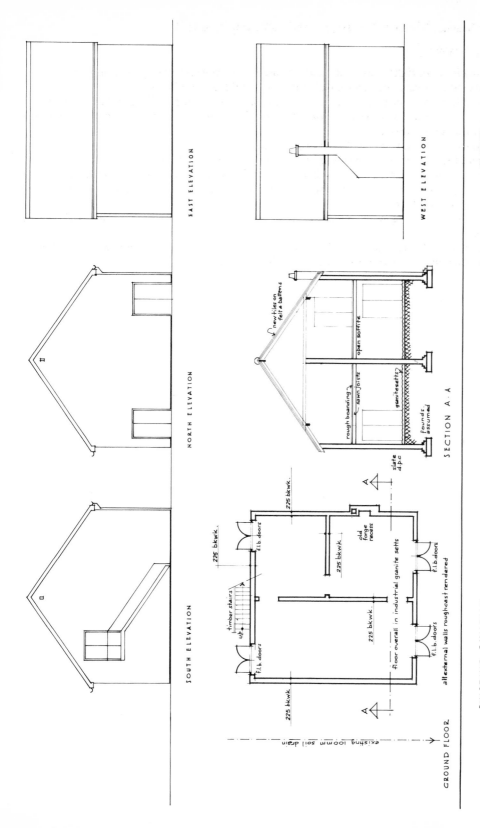

EAST ELEVATION

WEST ELEVATION

NORTH ELEVATION

SECTION A·A

new tiles on felt & battens

open soffite

rough boarding

sawn joists

granite setts

founds assumed

slate dpc

SOUTH ELEVATION

GROUND FLOOR

225 bkwk.

225 bkwk.

225 bkwk.

225 bkwk.

225 bkwk.

225 bkwk.

f.l.b doors

f.l.b doors

f.l.b doors

f.l.b doors

f.l.b doors

timber stairs

up

old forge recess

floor overall in industrial granite setts

all external walls roughcast rendered

existing 100mm soil drain

A

A

EXISTING BUILDING

OFB

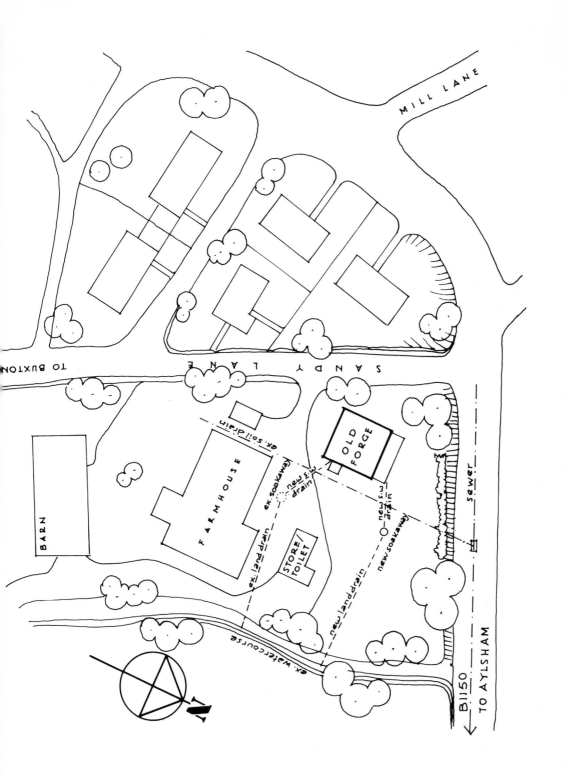

OFB 1

SITE PLAN

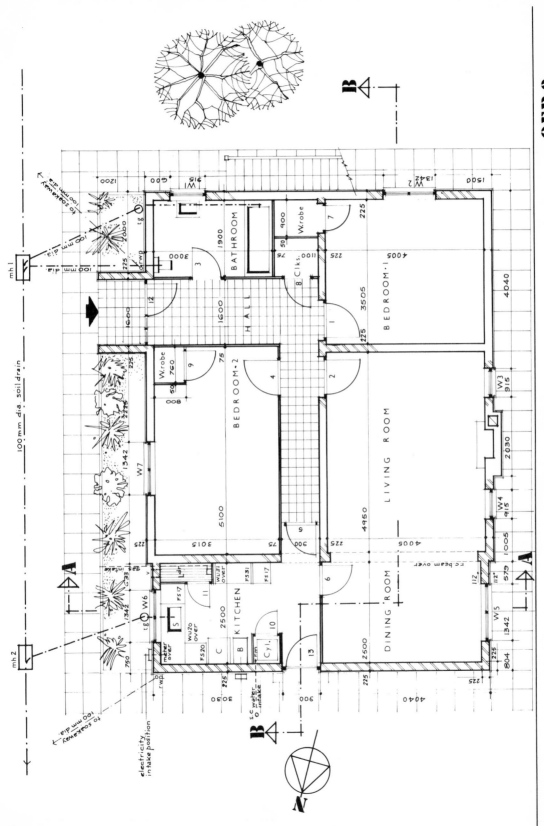

OFB 2

GROUND FLOOR PLAN

18

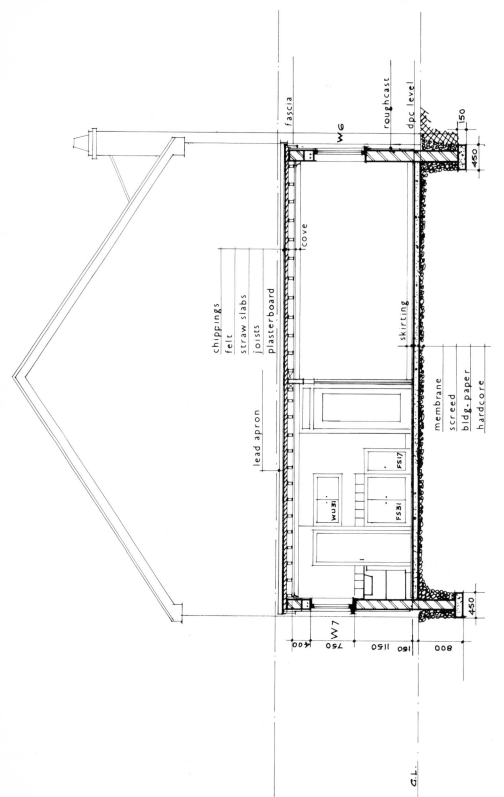

SECTION A · A

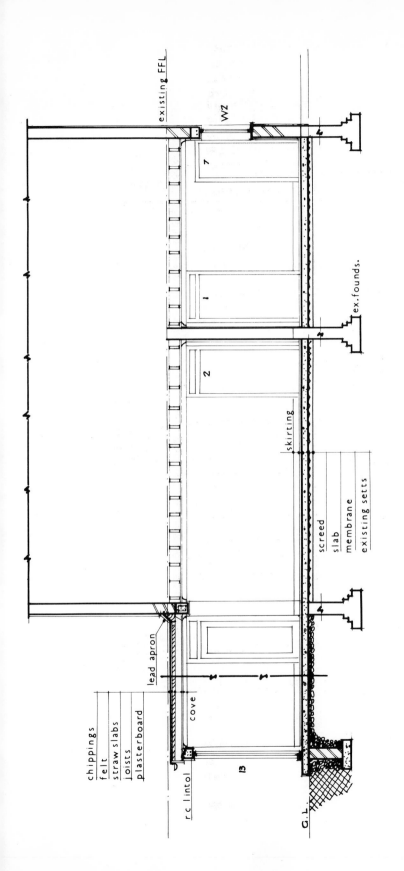

chippings
felt
straw slabs
joists
plasterboard

r c lintol

lead apron

cove

existing FFL

WZ

skirting

screed
slab
membrane
existing setts

ex.founds.

G.L.

OFB 3

S E C T I O N   B   B

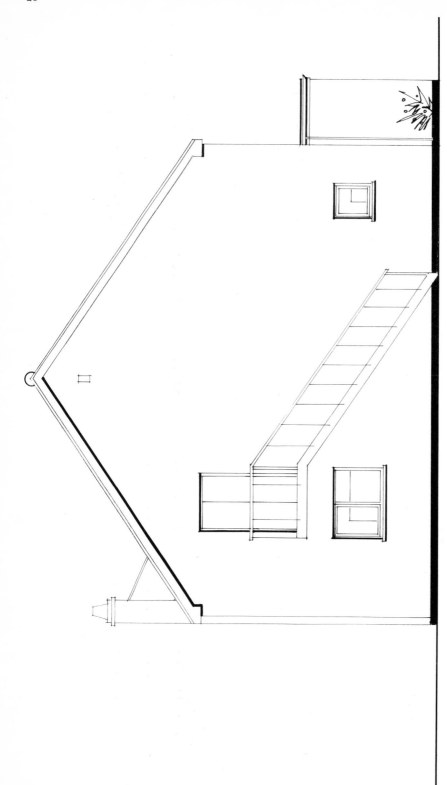

SOUTH ELEVATION

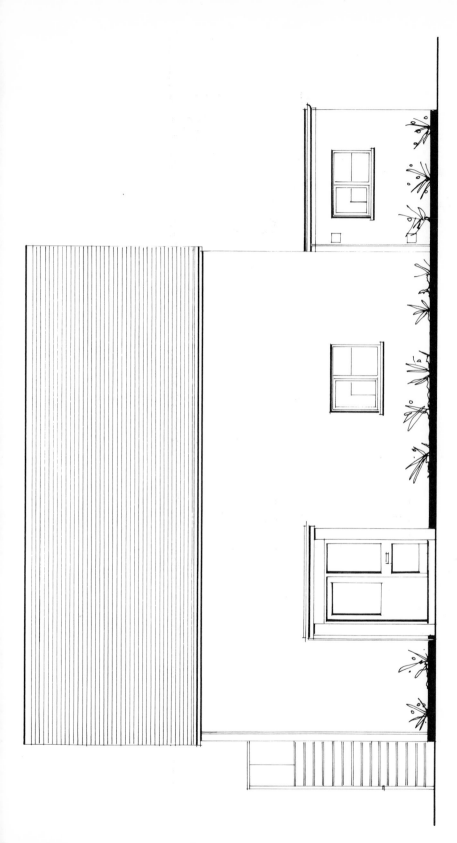

OFB 4

EAST ELEVATION

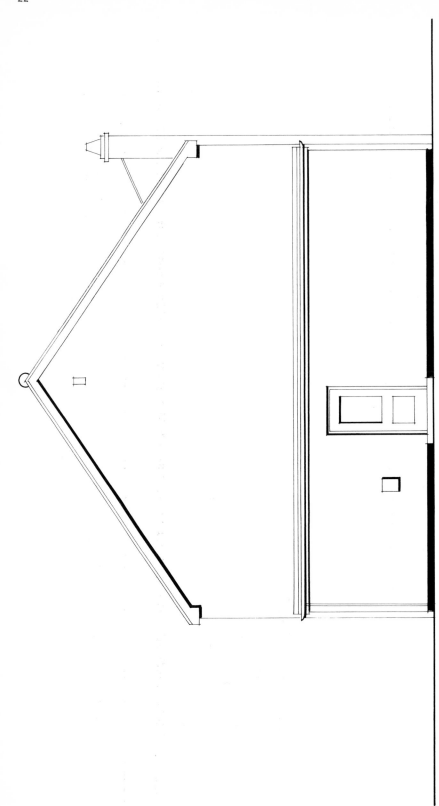

NORTH ELEVATION

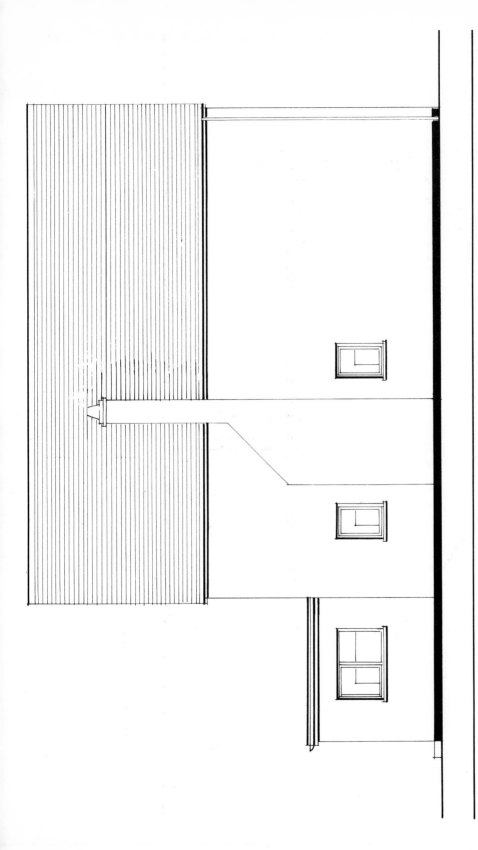

OFB 5

WEST ELEVATION

Specification check list for alterations and additions to Old Forge Building, Manor Farm, Frettingham, Norfolk

1 Clear debris and rubbish from G.F. and area adjacent to the building.
2 Strip site over the extension area and around the building perimeter.
3 Remove and store on site 4 No. pairs of double doors and frames.
4 Excavate extension foundations, form concrete footings.
5 Build brickwork to extension up to and including d.p.c. backfill.
6 Excavate manholes, drains. Form manholes, lay drains, gullies, etc., backfill.
7 Consolidate, lay hardcore over extension and porch areas.
8 D.p.m. over granite setts.
9 Concrete slab over extension and G.F. form porch slab and step to kitchen.
10 Remove existing rendering from north and east walls.
11 Build extension bk. wk. about d.p.c. level, build porch bk. wk. form entrance opening.
11a Remove existing brickwork north wall, brick up existing opening, form new piers.
12 Form openings for external window and door frames, cast lintels.
13 Build in external window and door frames.
14 Form openings for doors to Bedroom 1, Living Room, Wardrobe.
15 Cast beam over Living Room and Dining Room.
16 Work to existing forge opening.
17 Build internal partitions, form openings, build in frames, cast lintels.
18 Roof to extension and porch.
19 Windows and glazing.
20 Porch entrance frame.
21 Point windows and door frames.
22 Internal door frames.
23 Plastering.
24 Sundries in carpentry.
25 D.p.m. over slab.
26 Screeding.
27 Skirting and architraves.
28 Ceiling soffits and coves.
29 Glaze transom lights.
30 Doors and ironmongery.
31 Sanitary fittings and plumbing
32 Tiling.
33 Kitchen fittings.
34 Rendering externally.
35 Decorations internal and external.
36 Heating, hot and cold water supply.
37 Electrical installation.
38 Gas and water supply.
39 Heating, hot and cold water plumbing.
40 External plumbing.
41 Floor finishes.
42 Paving.
43 Flower beds.

Specification for alterations and additions
to
Old Forge Building, Manor Farm, Frettingham
for
A. Client

Preliminaries

1	Access to the site	The site is on the B1150 road 8 miles west of Aylsham and 7 miles east of Norwich. The site is entered from Sandy Lane from the main road. The lane is unmade, uneven, consisting of loose gravel and earth. Include for temporary crossover into the lane and support for wheel loads into the site from the main road.
2	Materials and goods storage	All goods will be stored and material tipped within the confines of the site area in positions to be agreed with the Employer.
3	Existing store and toilet	An existing store and toilet on the east boundary of the site will be made available to the Contractor throughout the contract period, who will maintain and clean same.
4	Site telephone	An existing overhead service is provided to the farmhouse on site. The Contractor to include for a site telephone to be installed throughout the contract period, and pay all charges in connection therewith.
5	Tree protection (see page 65)	Provide adequate protection to the trees at the site entrance 1 No. Chestnut 3 m girth, 1 No. Beech 2 m girth throughout the contract period.
6	Visit the site	The Contractor is to visit the site for the purposes of tendering, and no claim will be recognized in respect to want of knowledge of the site or its environs or the conditions thereof.
7	Contract	The conditions of the contract will be those as described in the Joint Contracts Tribunal Agreement for Minor Building Works (Without Quantities), 1980 current edition. A copy of the Form of Contract may be obtained from The Royal Institute of British Architects, 66, Portland Place, London W1N 4AD.
8	Drawings	The following listed drawings, together with the Specification, will be attached to the Form of Contract, and read together will form the contract documents. OFB 1, OFB 2, OFB 3, OFB 4, OFB 5 These drawings may be supplemented at the discretion of the Architect as provided for in Clause 5.4 of the contract.
9	Contractor's responsibility	The care of the works and all appertaining thereto is the responsibility of the Contractor, from the handing over of the site until final completion of the contract. The work of domestic and nominated sub-contractors is included in this overall responsibility. The Contractor shall provide everything necessary for the proper completion of the works in accordance with the terms of the

contract, and the true intent and meaning of the drawings and the specification read together.

The Contractor will provide all the necessary labour, plant, tools, etc., as required for the works, maintain same and remove from the site on completion of the contract.

The Contractor shall protect the property of all public utility and other service undertakers, and the local authority, as may obtain on or in relation to the site.

10	Attendance	*The Contractor shall attend upon and perform all builder's work necessary for and in all trades, including the work of all sub-contractors throughout the course of the contract. Unload, get into the site, all goods and materials until required in the works. Return all crates, containers, etc., carriage paid as required.* *The Contractor will allow all sub-contractors the use of water, lighting, plant, scaffolding, and other site facilities as required for the proper completion of their respective works.*
11	Water for works	*A mains service exists adjacent to the site, and the Contractor shall include for the provision of a supply of clean water for the works, install all temporary piping, draw off points, meters, etc., as required, maintain same during the progress of the works and dismantle and remove same on completion of the contract and pay all charges in connection therewith.*
12	Temporary power and lighting	*A single-phase service exists on the site and a three-phase supply to a transformer 110 m west of the site on the main road.* *The Contractor shall include for all temporary power and lighting required during the progress of the works, install and maintain same and dismantle and remove on completion of the contract and pay all charges in connection therewith.*
13	Workmanship	*The works shall be carried out by skilled tradesmen, unskilled men to be employed on labouring work only. All work shall be executed in accordance with accepted building trades practice to the satisfaction of the Architect. Materials as specified may be varied only with the written consent of the Architect. Samples requested by the Architect shall be submitted free of charge.* *The term BS shall mean the current British Standard Specification of the British Standards Institution, 2 Park Street, London W1A 2BS. The term BSCP shall mean the current Code of Practice of the British Standards Institution.*
14	Person in charge	*The Contractor shall notify the Architect in writing at the commencement of the works of the name of the person in charge of the works as provided for in Clause 10 of the contract, and of any changes which may take place during the progress of the works.*
15	Site facilities	*The Contractor shall provide such site facilities and accommodation for his operatives, including those of all sub-contractors, in accordance with the Construction (Working Places) Regulations 1967. Such accommodation provided shall be maintained during the course of the contract and removed on completion.*

16	Statutory Notices	The Contractor will provide the Architect with a copy of all Statutory Notices in relation to the works in accordance with Clause 6 of the contract.
17	Sub-contractors	All nominated and domestic sub-contractors shall be employed upon the works strictly in accordance with the terms of the contract, and the Contractor shall comply with all the statutory procedures required by any Act of Parliament.
18	Insurance	The Contractor is responsible for all insurances in addition to those required under the terms of the contract, under National Health Insurance, Sick Benefit, Unemployment, Holidays with Pay Schemes, Employer's Liability, Fatal Accidents, Workmen's Compensation, or any amendments thereto coming into force during the course of the contract, or against any claim at common law arising from the works.
19	Variations and Dayworks	Variations and Dayworks as provided for in the contract shall not be carried out by the Contractor or any sub-contractor without the written authority of the Architect.
20	Clear rubbish	Clear away from the site and adjacent areas all rubbish and debris to tip during the course of the contract and on completion. All access ways to the site and adjacent access roads to be kept free from mud and debris during the progress of the works. All materials dumped or stored on the site shall be contained and protected, and lighted where they may be a hazard.
21	Clean on completion	On completion of the works and immediately prior to handing over the works to the Employer, the Contractor shall be responsible for the cleaning of all floors, pavings, internal and external glass surfaces, removing of stains, paint spots, and other blemishes from fittings and other surfaces. Removing and taking away all protective coverings, etc., signboards and hoardings, and leaving the premises fit for occupation by the Employer to the satisfaction of the Architect.

Preambles to trades

Excavation

Hardcore — Shall be clean, dry, broken brick or stone, free from all deleterious matter and graded 150/50 mm sieve. Lay in layers not exceeding 225 mm at any one time, ram and consolidate to reduce levels as required.

Backfill — Shall be hardcore a.b.d. for paved areas or areas subject to wheel loads, or otherwise as specified in the works, consolidate a.b.d.

Make up levels — Levels where over-excavated shall be made up at the Contractor's expense as follows: Under load bearing areas in mass concrete 1 : 10 mix cement and mixed aggregate a.l.d. Under non-load bearing areas in hardcore a.b.d.

Plank and strut — Supply and fix all timber planking and strutting to support excavations until backfilled, strike and remove as backfill proceeds.

Removal of water — Keep excavations free from water by pumping or baling. Dispose of water so as to avoid nuisance to adjoining properties.

Soakaways — Excavate and fill as described in the works (page 70).

Concrete

Cement — Shall be Ordinary Portland Cement to BS 12.

Aggregates — For mass concrete shall be Thames ballast to BS 882.
Fine aggregate for reinforced concrete shall be clean, sharp, pit or river sand to BS 882.
Coarse aggregate for all concrete shall be crushed gravel to BS 882 graded to 18 mm dia.
'All in' aggregate shall be fine and coarse aggregate ready mixed to BS 882.

Water — Shall be clean (drinkable) and free from all organic impurities and at a temperature of 5° C and rising.

Reinforcement — Shall be round section mild steel bars to BS 4449 free from all loose mill scale, rust and grease, and wire brushed before placing in the formwork. Steel shall be cut and bent as shown on the bending schedules, or as described in the specification or shown on the drawings. Place in the formwork as shown on the drawings with spacing to comply with BSCP 112 'Reinforced Concrete' and tied with galvanized iron wire.

Formwork — Shall be well seasoned sawn softwood, properly secured to retain concrete without movement or distortion. Strike in accordance with BSCP 112 recommendations.

Mixing — Materials for mass concrete to be measured by volume in batch boxes.
Materials for reinforced concrete to be measured by weight.
Mix A 1 : 10 cement and coarse aggregate a.b.d.
Mix B 1 : 6 cement and 'all in' aggregate a.b.d.
Mix C 1 : 2 : 4 cement, fine aggregate, coarse aggregate a.b.d.

shown on the drawings in accordance with the details recommended in *CP 144 Part 2*.

| 7.5 | Tanking | * | Form asphalt tanking as shown on the drawings in mastic asphalt a.b.d. in three coats to a total thickness of 18 mm to walls and three coats to a total thickness of 28 mm to floors laid breaking joint. Form two coat angle fillet a.b.d. at all junctions of horizontal, vertical, and sloping surfaces. |

Form asphalt tanking as shown on the drawings in mastic asphalt a.b.d. in three coats to a total thickness of 18 mm to walls and three coats to a total thickness of 28 mm to floors laid breaking joint. Form two coat angle fillet a.b.d. at all junctions of horizontal, vertical, and sloping surfaces.

All brickwork joints to be raked to a depth of 19 mm for key, concrete surfaces to be roughened by hacking or by a slurry of sharp sand and cement thrown on to the surface.

Pipes passing through the tanking to receive a three-coat sleeve of asphalt prior to building in and two-coat angle fillets to be formed around the pipe at the junction with tanking after placing in position (see CP 144 Part 2).

7.6   Protect asphalt

The Main Contractor shall case up, cover, and adequately protect all asphalt work from excessive heat, abrasion and impact, the action of alkalis, acids, oils or other solvents, and deliver up the asphalt in clean and sound condition.

Paving

Introduction

The section which follows deals with some of the types of pavings in general use, alternatives have been included in addition to the specified works.

Wet screeding processes such as magnesite, terrazzo, granolithic, and epoxy and other resinous screeds are usually put in the plastering trade. Similarly, the dry flooring types such as cork, linoleum, rubber, timber, sheet and tile vinyl, carpet, etc. are often taken out as a separate description under 'Paving', or simply 'Flooring'.

Many firms undertake comprehensive flooring contracts on a specialist basis and would be dealt with as a P.C. sum.

It is important that the correct maintenance procedures are used and made known to the occupiers of the building at the time of handing over on completion of the contract. Some firms provide a brochure for this purpose, and may offer continuing maintenance contracts.

Paving

7.7	*Materials*	Sand:	*As described for Concrete.*
		Cement:	*As described for Concrete*
		Mortar:	*As described for Brickwork.*
	* Bricks:	*For paving to be reclaimed first quality stock bricks thoroughly cleaned before re-use.*
			or
	*		*Blue (or red) Staffordshire engineering bricks.*
	* Flints:	*To be selected smooth, sea-shore cobbled flints well washed before laying.*
		Paving slabs:	*To be precast concrete plain. Colour to BS 368 900 mm × 600 mm × 50 mm (or 62 mm) thick.*

Note: Coloured slabs to BS 1014 coloured right through or surface of the slab only to be stated in the specification. Paving slabs in pedestrian and vehicle areas to be 62 mm thick.

7.8	*Subsoil preparation*	*After removal of the top soil the area to be paved shall be spread and levelled with 50 mm thickness of crushed gravel consolidated to the required level with a 350 kg vibrating roller or a 2·5 Mg smooth roller.* *A weed killer used in accordance with the manufacturer's directions to be sprinkled over the area prior to laying the slabs.*

Note: This treatment may be omitted depending upon the nature of the sub-grade to be paved.
Where the subsoil is non-cohesive, e.g. ballast, the fill material may be omitted and the compaction only carried.

7.9	*Laying of slabs*	*Lay the slabs on 27–37 mm of lime, sand, mortar 1: 3 mix by volume, spread evenly over the entire area of the slab as laying proceeds. The joints to be broken (or such other pattern as may be required) and slurried with a mix of mortar a.b.d., the residual slurry removed with a squeegee.*

Note: In urban areas joints should be as close as possible, but open joints are satisfactory if properly filled and rammed with cement mortar. In rural areas slabs should be 'buttered' on all edges with lime/sand mortar 1: 3 mix a.b.d. when laid and all joints filled with cement/sand mortar 1: 3 mix a.b.d. 5 mm wide to prevent weed growth. For light traffic areas only.
Procedure for crazy paving is as described for slabs but the slabs broken into irregular shapes, or may be irregular shaped York Stone. The bedding may be varied to suit the conditions, but 37–50 mm clean sharp sand if the subsoil is firm.

8.0	*Tamping*	*All slabs to be laid firmly on the bed and tamped into position with a pavior's maul, care being taken to ensure slabs do not rock.*

8.1	*Cutting and waste*		*Allow for all straight, raking, and circular cutting as required including for all waste.*
8.2	*Cobbles and flints*	*	*Prepare subsoil and bedding all as described for slabs.* *Lay 75 mm thick bed of concrete 1 : 2½ : 4 mix of cement/sand/fine aggregate by volume spread roughly level, lay cobbles or flints by hand to half their depth in mortar mix A a.b.d. as soon as possible after the base is laid. After setting sprinkle with water and brush hard to remove all residual mortar.*
8.3	*Bricks*	*	*Prepare sub-soil and bedding all as described for slabs using only semi-engineering or engineering bricks (hard burnt paviors). Lay bricks to pattern on an 18 mm thick mortar bed a.b.d., tamp bricks level, or to face as required.*

Note: Bricks can be laid on a sand bed, but may become uneven with wear.
Joints can be flushed-up and pointed, or left open, or filled with sand.
Combinations of brick, flints, cobbles, etc. can provide interesting patterns.
Soft bricks will suffer frost damage, if used.

| 8.4 | *Kerbs and edges* | * | *Abutting paths and vegetation areas:* *To be edging strips in precast concrete to BS 340 laid to required falls and bedded on 100 mm of concrete mix B a.b.d.* *Abutting roads:* *To be 250 mm × 100 mm battered kerbs with rounded top arris laid and pointed a.b.d.* |

Note: *In-situ* concrete paths and driveways can be treated with carborundum dust 454 g per square metre sprinkled over the surface during tamping, which will provide a reasonably non-slip surface but will affect the colour of the concrete, mica can be used in a similar way, but will only provide some sparkle to the concrete.

Roofing

Introduction

What is described under this trade is only a guide and does not attempt to cover every type of roofing of the slate and tile variety. The more common types are described, including cedar shingles, all based upon minimum construction, with other alternatives given to improve the basic construction for better class work.

Where pantiles or other single or double lap tiles are used in lieu of plain tiles, the procedure in the description is the same, but with variations in battens and gauge.

Roofing works may also be specified on a P.C. sum basis. This procedure is usual where specialist copper and other types of built-up proprietary roofing systems are used. Repairs and specialist treatments for the waterproofing of existing roofs e.g. 'Turnerizing', would also be dealt with in this way.

Be sure that all builder's work has been included in specifications. Points to check are tank bearers and access boards, access traps, tank covers, lagging, insulation, flashings, eaves filling and soffits, barge boards and soffits, metal trims, etc. Make a thorough check list of all items to commence, especially in renovation work, and check it against specialist estimates or specifications. If in doubt include a provisional sum to cover additional work or describe it.

Roofing

8.5	*Tiles*	*Shall comply with BS 402 and shall be hand made, sand faced, cambered plain tiles with nibs, twice drilled for nailing. Average size 250 mm × 160 mm.*
	*	*alternatively*
		As above, a.b.d. but machine made.
	*	*alternatively*
		As above, but concrete to BS 473.
		All tiles to be equal in all respects to samples approved by the architect.
	*	*Where valleys or hips are used in the roof construction.*
	*	*Valley tiles shall be purpose made swept valley tiles to match general tiles a.b.d.*
		alternatively
		Plain tiles cut as required with tile and a half to bond.
		Hip and ridge tiles
		Ridge tiles shall be half-round pattern to match general tiles for texture and colour 450 mm × 200 mm bedded and pointed in mortar mix A a.b.d. with ends solidly filled with slip tile inserts.
		Hip tiles shall be as described for ridge tiles with ends supported on black japanned steel hip irons nailed to hip rafter and set in mortar a.b.d.
8.6	*Nails*	*Nails for battens shall be 42 mm gal. steel wire nails.*
		Nails for tiles shall be 42 mm copper or alloy nails.
8.7	*Under-eaves and ridge tiles*	*Lay a double course of tiles at the eaves and the ridge, undertiles to be 175 mm × 160 mm. Bed solidly in cement mortar a.b.d. one course of plain tiles. Over gable walls at verges under tiled slopes to project 37 mm beyond the wall face and point in mortar a.b.d.*
8.8	*Laying of tiles*	*Gauge:* *Shall be 100 mm c/c of battens with extra battens for under-eaves and ridge tiles.*
		Margin: *Shall be 75 mm*
		Lap: *Shall be 62 mm*
8.9	*Workmanship*	*Tiles shall be laid by skilled craftsmen.*
		Lay over the entire roof slopes over the rafters two-ply roofing felt to BS 747 Class 1 (a) laid with 300 mm laps and dressed over the ridge and tilting fillet, secure with 42 mm × 42 mm sawn s.w. battens laid to gauge a.b.d. and nailed to every rafter. Lay tiles breaking joint double lapped in even courses vertically and horizontally with a tile and a half in each alternate course to break
	*	*the joint. Lay under the ridge, eaves, and verge tiles a.b.d. Form valleys and hips a.b.d. allowing for all cutting and waste.*
		Note: If soakers are required, either lead, zinc, or aluminium, these are to be described in 'Plumbing' external works and described as 'fix only' in 'Roofing'. Similarly with metal ridges, hips and valleys.
		Leave roof sound and watertight on completion.

The following are alternatives for different types of tiles:

* *Pantiles: Clay, concrete, or glazed clay*
Gauge: *260 mm*
End lap: *75 mm*
Side lap: *45 mm*
Battens: *50 mm × 25 mm*
Each tile nailed once. One course of plain tiles at eaves. Ends of eaves tiles filled with mortar mix A. a.b.d. Tile inserts bedded in mortar at the ridge.

* *Single and double Roman tiles:*
Gauge: *250 mm*
End lap: *75 mm*
Battens: *50 mm × 25 mm*
Each tile nailed twice.
One course of plain tiles at eaves. End of tiles at eaves filled with mortar mix A a.b.d. Tile inserts bedded in mortar at the ridge.

* *Flat interlocking—Roman interlocking tiles.*
Gauge: *260mm*
Battens: *50 mm × 25 mm*
Each tile nailed twice.
One course of plain tiles at eaves. Ends of tiles filled with mortar mix A a.b. Tile insets bedded in mortar at the ridge.

* *Spanish tiles:*
Gauge: Vertical battens 75 mm × 50 mm at 260 mm c/s
Roof boarding sawn s.w. T & G or plain edged.
End lap: *75 mm*
Ends of tiles filled with mortar mix A a.b.d. and tile slips.
Channel tile side nailed, cover tile top nailed to battens.
One course of plain tiles at eaves.
Plain tiles cut and set in mortar at the ridge.

* *Western Red Cedar shingles:*
Gauge: *125 mm (varies with pitch)*
Lap: *150 mm*
Margin: *125 mm*
Battens: *50 mm × 25 mm or 45 mm × 18 mm*
Counter battens: *450 mm × 600 mm*
Nails copper or alloy a.b.d.
Each shingle nailed twice with 4 mm clearance between shingles to allow for swelling.
Double course of shingles at eaves.

* *Slates: Wales, Cornwall, Scotland, Westmorland, Belgium*
Gauge: $\dfrac{length\ of\ slate - lap}{2}$

e.g. $\dfrac{450 - 75 = 188\,mm}{2}$

Common sizes: *600 mm × 300 mm*
                  500 mm × 250 mm
                  450 mm × 225 mm
                  400 mm × 200 mm

Battens and counter battens 50 mm × 18 mm
Nails equal to twice slate thickness + 25 mm copper or composition
nails (never galvanized steel)
Boarding plain edge or T & G in good class work.
Double courses of slates at eaves, ridge, verges, slate and a half at
hips and valleys.

Note: All the foregoing would have roofing felt included
a.b.d. for tiling. Some cheaper work may use
building paper in lieu but this is not
recommended.

8.10	*Built up felt roofing*	

Material:
Shall be heavy grade to BS 747 Class 1 (a) (or otherwise as
required).
Workmanship shall comply with BSCP 144, 101.
Felt to be applied in three layers laid breaking joint, each layer
bonded with hot bitumen, and rolled to exclude all entrapped air.
Form internal angle fillet and 75 mm upstand at all abutments and
dress into brick joint (or carry through as d.p.c. as required). Lap
and mitre felt at all internal and external angles.

Note: Metal apron flashings to be described under
'Plumbing' external work. Bedding and pointing
in 'Brickwork'

* Eaves finish varies with the job in hand:

Dress over fascia board with chamfered edge and form 50 mm deep
single lap welt around all edges, lap and mitre at all corners.
<u>*alternatively*</u>
Dress felt 25 mm over fascia board a.b.d. and fix aluminium trim
(as described in Metalwork) over felt edge and fascia board and
secure with aluminium screws at 450 mm c/s.
* *Drips and outlets:*
Cut and form felt into rainwater outlets, around pipes as required
in accordance with BSCP 144.101 details.
* *Finishes vary with the job in hand:*
To be left plain finished.
<u>*alternatively*</u>
The surface to be finished with a layer of fine white sand (or spar
chippings) rolled in cold bitumen.

Note: Roof coverings of the waterproof membrane type,
e.g. atactic polypropylene (APP), plain or with
slate granule finishes are alternatives to felt. Some
carry ten-year guarantees

Carpentry and joinery

Introduction

Although these two trades are usually linked together, and one hears the Tradesman described as a Carpenter and Joiner, they are really quite separate trades. Carpentry is concerned with framing and carcassing construction where things are nailed together rather than screwed, although in some instances the edges of the trades become blurred, and on site these Tradesmen will do each other's work on occasion.

In Painting and Decorating reference is made to the present nature and characteristics of softwood for use in carpentry and joinery work, over which, with the best will in the world, the Contractor has little control in terms of quality. It is a constant problem for the industry as a whole, with the use of young growth, kiln-dried timber, pink, resinous, soft sapwood, which produces the characteristic warping and twisting, and distortion when drying out, especially in centrally heated areas.

There is no real solution to the problem, one can only reject on site the really poor quality timber and hope the replacement will be better. One has only to observe the quality of timber sold in DIY shops to see the extent of the problem.

Hardwoods, with the influx since the end of the war of African and other imported hardwoods, have become difficult to select. There are inherent problems in these timbers, such as gum bubbles, difficult grains, and other defects which manifest themselves in use. There is a lot of useful information about them and other hardwoods available in BS 881, BS 1186 and BS 589, and from the Timber Research and Development Association. In specifying door frames, linings, and doors, check that they will be compatible with the fire resistance requirements for the position in which they are to be fixed.

Make sure when specifying standard fitments of any kind, that provision for fixing is included, and the same with sanitary fittings, especially where demountable partitions are used, as they require special inserts for this purpose.

Ordinary 'stud' partitions are seldom met with in new works, but often in rehabilitation and conversion works. In some period buildings they may be structural 'trussed' partitions providing support for the carcass of the building. The specifier must be careful in describing openings to be formed in them, or their removal, as serious problems can ensue if they are interfered with without due regard for their structural function.

Formwork for reinforced concrete comes within this trade, but is described under Concrete, and some Carpenters become familiar with this type of work and do nothing else.

Carpentry and joinery

9.0	Materials	Timber: Carpentry work

Shall be sound, bright, Douglas Fir, free from large, loose, or dead knots and waney edges. The timber shall comply in all respects with BS 1186, Part 1, and have a moisture content not exceeding 20% dry weight.

Timber: Joinery work

Shall be European Redwood, equal in standard to that classed as 'clears and door stock', and shall comply in all respects with BS 1186, Part 1. The moisture content shall be relative to the position in which the timber is to be fixed, and between the limits of 8–14% dry weight when fixed internally, and 15–17% dry weight when fixed externally.

Timber for the works including all made-up joinery shall be stored under cover in dry, warm conditions and protected against damage before and after fixing.

Sizes:

Where sizes are shown on the drawings as 'ex' an allowance of 3 mm shall be deducted for each prepared face in softwood. For hardwoods the deduction shall be 1·5 mm for each prepared face.

Plywood:

Plywood for internal use shall comply with BS 1455 and face graded 2nd quality B/B.

Plywood for external use shall comply with BS 1455 and the adhesive used in manufacture to comply with BS 1203 WBP (this can also be BP or MR). The edges of all external plywood to be treated with two coats of bitumastic paint before building into rebates.

Nails:

Shall comply in all respects with BS 1202.

Screws:

Shall comply in all respects with BS 1210 and shall match the metal to be fixed, i.e. brass to brass, steel to steel, etc., except as otherwise specified.

Preservatives:

Softwood for carcassing and joinery work shall be treated with a water-borne preservative to BS 4072. The ends of all cut timbers shall be steeped in preservative on site before fixing. All timbers to be built in to be immersed in preservative a.b.d. for 24 hours before building in. When brush applied, three full coats shall be applied at 15-minute intervals.

9.1	Adhesives	*All internal joinery shall be put together with a good quality joiner's glue or a proprietary synthetic resin adhesive to BS 1204 Part 1 (INT).*

All external joinery shall be put together with synthetic resin adhesive complying with BS 1203 of not less grade than BP.

9.2	Workmanship	*All carpentry work will include for all labours and materials required in framing up and carcassing, rough grounds where required for fixing shall be treated with preservative to BS 4072.*

Joinery shall be put together with all necessary tenoning, morticing, grooving and tonguing, rebating and housing, and other labours necessary for the proper completion of the work.
The joiner shall supply and fix all necessary plates, brackets, nails and screws, required for completion of the works.
Windows and door frames will be built in as the carcassing work proceeds and properly squared up and braced, and secured with 4 No. per frame, once bent, twice drilled, galvanized steel cramps ragged tailed for building into brick and blockwork, and screwed to frames with alloy screws.

9.3	Suspended first floor	*Carcass the floor with 175 mm × 50 mm joists at 375 mm c/s supported on gal. steel joist hangers built into the bkwk. a.b.d. Strut through the centre of each span with 50 mm × 37 mm sawn s.w. herringbone strutting.*

Under demountable partitions as later described fix double joists joined together with steel gang nail plates top and bottom at 450 mm c/s staggered.
Where partitions run against the span solid bridge under the partitions with 150 mm × 75 mm solid sawn s.w. bridging skew nailed to joists.
Cover the floor with 125 mm × 25 mm nominal T & G wrot. s.w. boarding cramped together, laid with heading joints staggered and twice nailed to joists. Allow for expansion around the perimeter of all boarded areas, punch nail heads.

9.4	Pitched roof	*Carcass the roof with SNL gang nail trusses types 720 and 725 spaced at 600 mm c/s secured to 100 mm × 50 mm sawn Fir plates, each truss tied at each end with 25 mm wide 250 mm long gal. steel straps. Secure wallplate with 300 mm × 25 mm flat bar gal. steel straps once bent over plate and secured to blockwork.*

Trusses to be braced in accordance with Technical Bulletin Sheet No. 1 of The International Truss Plate Association.
Fix water tank support and access boards in accordance with the Bulletin last described, Sheet No. 4.
Eaves overhang dimension to be 225 mm. The eaves soffit boards to be external quality plywood a.b.d. fixed to wrot. s.w. battens secured to bkwk. and housed into the fascia board.
Fix to the eaves 200 mm × 25 mm wrot. s.w. fascia board housed to receive soffit board a.b.d.
Fix along the length of the eaves at the foot of the trusses 175 mm × 50 mm wrot. s.w. tilting fillet.
Cover the roof slopes with nominal 100 mm × 25 mm sawn softwood boarding nailed to joists, level surface to receive felt and battens as described in 'Roofing'.
Fix the gable ladders as supplied with trusses according to the manufacturer's instructions.
Fix to the gable ladders 175 mm × 25 mm wrot. s.w. barge boards with shaped ends to finish flush with the bottom of the fascia boards, and housed to receive soffit boards a.b.d.

9.5	Weatherboarding	*Fix to the external walls as shown on the drawings 110 mm × 16 mm wrot. Cedar wood boards rebated and 'V' jointed secured through a layer of building paper to 25 mm × 25 mm sawn s.w.*

battens nailed to blockwork with non-ferrous nails.
Perform all labours in cutting to openings, gable ends, etc., all external angles to be lapped jointed.
Finish boarding with sealer as described in external decorating.

| 9.6 | Demountable partitions | *Fix in the positions as shown on the drawings demountable partitions to comply with BS 4022 'Prefabricated Gypsum Wall Panels, Type 1'. Provide all timber plugs for fixing a.b.d. Perform all labours in cutting and fitting to sizes required. Form all openings as shown and fix frames and linings as later described.* |

| 9.7 | Trim staircase opening | *Trim the staircase opening as shown on the drawings with 150 mm × 75 mm fir trimmer joists.* |

| 9.8 | Staircase | *Construct the staircase from ground to first floor as shown on the drawings as follows:*
In wrot. s.w. throughout a.b.d. 825 mm in clear between strings with 30 mm finished thickness treads and winders with rounded nosings, 25 mm finished thickness risers, all tongued and grooved together, glued and blocked, and bracketed on 125 mm × 75 mm carriage piece framed to trimmer at first floor landing and to bearer at the bottom of staircase. House and wedge tread and risers to wall strings and glue and block. Strings to be 275 mm × 35 mm wrot. s.w. blocked out and plugged and screwed to walls 18 mm clear of face of blockwork. Strings to be splayed 10° on top edge with arris rounded.
Handrail to be 50 mm dia. round section with flat for core rail and rounded ends, in Agba hardwood bodied in and french polished. Fix rail with 4 No. stainless steel offset brackets with 30 mm flat stainless steel core rail drilled and c.snk. at 600 mm c/s. Plug and screw brackets to wall 875 mm above the string with stainless steel screws provided. Screw handrail to core rail.
Form the cupboard under the staircase with 75 mm × 50 mm sawn s.w. studs nailed to sole plate, strings, and carriage piece. Cover the studding and staircase soffit with plasterboard, scrim and set as described in Plastering.
Form the door opening under the staircase and fit lining, door, and ironmongery as described in the Window and Door Schedule. Fix architraves b.s. as described.
Allow the Provisional Sum of £X.00 (amount in words) for wrot. Hardwood protection cappings to returns of partitions at ground and first floor staircase levels. |

| 9.9 | Skirtings | *Fix around the perimeter of all rooms, hall, landing and passages, 100 mm × 18 mm wrot. s.w. splayed and rounded skirtings. Prime on all faces before fixings.* |

| 10.0 | Doors, frames, linings, furniture, ironmongery | *Fix in the positions shown on the drawings the door frames, linings, doors, furniture and ironmongery described in the Door and Window Schedule.* |

| 10.1 | Windows | *Fix the windows in the positions as numbered on the drawings in accordance with the Door and Window Schedule. Frames in* |

*brickwork to be set 50 mm from external face of brickwork.
Windows in weatherboard cladding to have WSM surrounds as
shown in the Schedule.*

10.2	*Glazed screen to living room and front door*	*Make up and fix in the positions shown on the drawings in wrot. Afrormosia hardwood the glazed screen wall to the living room as follows:* *Overall size of frame 3000 mm wide × 2100 mm high. Head, sole, door, and side frames to be ex. 100 mm × 75 mm section rebated 18 mm × 18 mm for glazing and 40 mm × 18 mm for door and splayed from rebate to front of frame 12° with exposed arrises rounded.* *Make 3 No. mullions equidistant in the side framing as shown but twice rebated for glazing and twice splayed and rounded all a.b.d.* *Make the door in Afrormosia a.b.d. to suit the opening size × 40 mm finished thickness with 150 mm top and bottom rails and 100 mm side rails rebated 18 mm × 18 mm for glazing. Hang the door on a pair of 80 mm brass butt hinges.* *Make Afrormosia wrot. glazing beads ex. 20 mm × 20 mm splayed 5° on one face only cut to size with mitred ends and drilled and countersunk equidistant in the lengths for brass cups and screw fixing.* *Fix the frame in the opening with suitable rough grounds, centrally in the partition thickness, and secure with 6 No. frame cramps a.b.d. 3 No. to each end frame set in blockwork a.b.d.* *Fix round the perimeter of the frame b.s. ex. 50 mm × 18 mm wrot. Afrormosia hardwood architrave rounded on exposed arrises. Punch down all pin heads and fill with matching wood filler and rub down. Fix to the door a.b.d. 2-lever mortice lock complete with lacquered brass lever furniture with integral finger plates.*
10.3	*Front door screen*	*Make the screen and door for the front entrance all a.b.d. but with 1 No. mullion and with an ex. 150 mm × 62 mm rebated, weathered and throated Afrormosia threshold set on a metal water bar in mastic. Fix to the door 3-lever mortice night latch complete with lacquered brass lever furniture all a.b.d. and a brass letter flap.*
10.4	*Plant troughs*	*Include the P.C. sum of £X.00 (amount in words) for 2 No. plant troughs for the living room and front door screens a.b.d. obtained from Flora Designs Ltd, Aylsham, Norfolk. Catalogue Nos. 347 and 349. Allow for profit.* *Fix the trough in the living room to the screen mullions at 750 mm from FFL in 3 No. aluminium brackets as supplied by the manufacturers.* *Similarly fix the trough to the front door screen but 225 mm from FFL of porch.*

Note: Sundries in carpentry for shelving, hanging rails in
wardrobes and cloaks, cylinder supports etc., not
repeated in this section, see page 39 'spot item'
specification.

10.5	*Servery*	*Form the opening for the serving hatch between the kitchen and dining room to receive 920 mm × 460 mm × 30 mm wrot. s.w. lining with 40 mm × 10 mm wrot. s.w. planted stop. Head of opening 1400 mm from FFL.* *Prime all timber before fixing.* *Hang to lining on pair of 40 mm brass butt hinges a pair of 28 mm finished thickness louvred pine doors. Fix to top rail of both doors a brass magnetic catch. Fix centrally in both meeting rails a pair of 25 mm dia. brass knobs.* *Fix to lining b.s. primed wrot. s.w. 30 mm × 10 mm architrave with rounded arrises.*
10.6	*Roof space access*	*Trim with 100 mm × 50 mm sawn s.w. noggings the access to the roof space in the position shown on the drawings. Form the opening with 50 mm × 25 mm wrot. s.w. nett size 600 mm × 600 mm lining. Fit to the lining 25 mm thick blockboard access cover. Prime and paint a.b.d.*

Plumbing

Introduction

In a basic book on specification writing, this trade can be dealt with in only the simplest of terms, covering as it does such a wide variety of materials and installations.

Alternatives are given in the following section, which, together with the Plumbing section in the 'spot item' and the Preambles to Trades, will provide a basis for comparison and selection.

Plumbing

10.7	Materials	Soil pipes:

Soil pipes:

Will be 100 mm internal dia. cast iron with spigot and socket joints to BS 416 medium grade. Pipes to be fixed with C.I. holderbats in two sections bolted together, built into wall, wedged and pointed in cement mortar 1:3 mix. All bends, branches, junctions etc., will conform to BS 416. Joints will be molten lead run on to hemp gasket and well caulked.
Where carried up as ventilation the pipe will terminate 300 mm above the roof slope with an attached gal. wire balloon and a Class 4 lead flashing and collar around the pipe.

Waste pipes:
Will be 50 mm internal dia. C.I. to BS 416 fixed and jointed all a.l.d.

Rainwater pipes:
Will be 100 mm internal dia. or 75 mm internal dia. as required, medium weight socketed with ears cast on to comply with BS 460 fixed to walls with gal. iron distance pieces and rose-headed pipe nails with a clearance of 37 mm. Bends, offsets, junctions will be included where required and shoes where discharging over gullies. Joints to be made with tow and red lead putty.

Gutters:
To be 100 mm H.R. section C.I. to BS 1205 secured to fascias with C.I. brackets and gal. iron round-headed screws. Include for all stopped ends, outlets, as required.

* Alternatives:
Extruded aluminium	BS 1430
Galvanized mild steel	BS 1091
Wrought copper and zinc	BS 1431
Asbestos cement	BS 569

* Waste branches:
To be in lead to comply with BS 602 30 mm dia. for basins and 35 mm dia. for sinks, bath, and shower units.

* Jointing:
Joints to be wiped soldered complete with all brass ferrules, thimbles, etc., appropriate to the joint to be made.

Traps:
To be drawn lead to BS 504 30 mm dia. for basins, 35 mm dia. for sink, bath, shower wastes. Traps to have integral cleaning eye, brass tail pipe and coupling nut. To be 'P' or 'S' type as appropriate.

* Alternatives:
Copper and copper alloy all a.b.d. to BS 1184.

Overflow pipes:
Shall be drawn lead of suitable diameter to the outlet to discharge

into the open air 100 mm beyond the wall face and terminated with a chamfered drip.

* *Alternatives:*
Or other material as appropriate to the installation.

* *Cold water storage tank:*
Fix in the position shown on the drawings (or as specified) a galvanized steel tank of 750 litres actual capacity to BS 417 Grade A standard drilled for all connections complete with 'Croydon' pattern ball valve, arm, and copper ball.

Note: Header tank not specified, see page 33.

Note: Bearers, access boards, cover and top, insulation, should be as described in Carpentry and Joinery or Preambles to Trades.

Where lead is used for pipes, or galvanized steel or iron, a galvanized steel tank is in order to BS 417 Grade A a.b.d. Where copper piping is used, a plastic, asbestos, or copper tank should be used to avoid electrolytic action. The use of insulated washers, or painting the inside of the tank with bitumen compound is inferior and unreliable. This may be done to reduce chemical attack from the water in some areas so affected.

Support for pipes:
Lead pipes will be supported with gal. steel pipe clips of appropriate size at 750 mm c/s plugged and screwed to walls.

* *Anti-syphonage branches:*
To be drawn lead 25, 30, 37 mm internal dia. for basins, baths, and sinks respectively. Connections to be lead soldered not less than 75 mm and not more than 300 mm from the crown of the trap.

Cold water pipes:
To be lead to BS 602 No. 1 composition, Table 4.

Hot water pipes:
To be lead to BS 602 No. 1 composition, Table 4.
Pipes to be of suitable weights and bore sizes as required by the water supply authority.

Pipe runs:
All pipe runs to be neat and inconspicuous, where a drawing is not available, those exposed to view to be run in positions agreeable to the Architect.

10.8 | *Sanitary fittings*

Note: These may be dealt with as a P.C. sum as in the 'spot item' project or described in detail. The following may still be used in domestic work both new and rehab.

* Fix in the position shown on the drawings a C.I. 'Magna' pattern bath 1950 mm 675 mm over roll, with adjustable feet, all white glazed internally and over roll, complete with C.P. waste outlet, backnut, and union for lead (omit union for copper etc.), 37 mm drawn lead trap a.b.d. Pair of 'Hi-Flow' C.P. taps marked Hot and Cold, C.P. chain and attached plug. Connect cold and hot water services, waste, trap, and overflow a.b.d. Fix on sawn s.w. bearers plugged and screwed to walls and floor, 6 mm thick enamelled hardboard front panel (white) secured with C.P. dome-headed screws. Seal bath to wall with mastic.

108

*   Lavatory basin:
Fix in the position shown on the drawings a 'Lotus' pattern white vitreous china basin and pedestal to BS 1188, complete with pair of 'Hi-Flow' C.P. taps marked Hot and Cold, C.P. waste outlet, backnut, and union for lead, (omit union for copper etc.), chain and attached plug. Connect hot and cold water services, waste, trap, and overflow all a.b.d.

111

*   W.C. suite:
Fix in the position shown on the drawings a 'Lotus' pattern white vitreous china W.C. suite to BS 5503 complete with 'Lynx' pattern white enamelled steel w.w.p. with white enamelled steel flush pipe and cone connector to BS 5627, white plastic seat and cover and C.P. fixing lugs. Connect cold water service to w.w.p. complete with control valve a.b.d. Connect overflow pipe a.b.d. and m.g.

114

*   Shower trays:
Fix in the positions shown on the drawings 2 No. vitreous china shower trays 800 mm × 800 mm complete with integral overflow, C.P. waste outlet, backnut, and union for lead (omit union for copper etc.), C.P. thermostatic type mixer with temperature control and thermostatic valve, C.P. service pipe and adjustable shower head. Plug and screw shower head and mixer to tiled wall.
Connect hot and cold water service pipes a.b.d., trap and waste branch, overflow, a.b.d.

118

Note: First floor showers are a problem due to water seepage, and it is recommended that a purpose-made shower cubicle is used in such positions. Alternatively, a zinc or lead tray can be formed for the shower tray to sit in, and turned over a suitable height kerb, the cubicle tiled with ceramic tiles down to the kerb and sealed with a non-hardening mastic. With a shower curtain and impervious floor covering the water seepage can be contained.
Joists must not be cut to run the waste, the tray must be set above floor level to permit the waste to run above the floor surface. Hot water should be run from the top of the cylinder and the expansion pipe or loop not from another sanitary fitting service pipe. Cold water must not be at mains pressure but from the cold water storage tank, and if taken from another service pipe reduction in flow can cause scalding on some mixers.

10.9	Services, control valves, cylinder and boiler	See Plumbing description in 'spot item' specification page 33
11.0	Water storage heaters	Connect a.b.d. cold water service pipes to storage heaters described in Electrical Works.

11.1	*Rising main and water supply intake*	*See Plumbing in 'spot item' specification page 41.*
11.2	*External plumbing*	*Fix gutters, down pipes, complete with stop ends, outlets, in the positions as shown on the drawings all a.b.d.*
		Fit 300 mm × 300 mm lead flashing to BS 1178 Class 4 to all waste and soil and vent pipes complete with upstand collar.

Plastering

Introduction

The chart on page 123 shows the relationship between limes and plasters. In the plastering trade, the former is almost entirely used for gauging, and has been replaced by gypsum plaster for general work. The choice of plaster depends upon the environment in which it is to be applied, the compatibility of the backing surface, and the hardness of the finished coat. Some of the brands produced under the classes shown on the chart are as follows:

Class A   Plaster of Paris
Class B   Retard hemi-hydrate
           Adamant, Battle Axe, Gothite, Gypstone,
           Murite, Napco, Sirifix, Thistle
Class C   Anhydrous
           Sirapite, Statite, Victorite
Class D   Hard Burnt
           Keenes, Parian
Class EE   Anhydrite
           Pioneer

Special plasters such as Vermiculite and exfoliated mica plasters are used for their acoustical properties, but any applied decoration may destroy this, which limits their general application. Acoustic tiles may be treated with a factory-applied flame retardant such as Albi-R which is normally white.

The so-called 'bonding plasters' which are applied direct to concrete surfaces contain a matrix such as sharp sand or ground spar to increase the tenacity of the plaster.

Specify board plasters for plasterboard and plasterlath work.

Dry lining, although described under Plastering, may be done by others, and information on these special boards and their application can be obtained from the manufacturers, e.g. Gyproc Products Ltd.

Fibrous plaster work is done normally by specialist firms, but small amounts may be run on the site bench by the plasterer, and is used for decorative work such as cornices and other enrichments. The plaster used is Class A.

External renderings are included in Plastering and information on the many types and the specification of them is given in the Bibliography. Other processes included in this trade are floor screeding, covering sand and cement, granolithic, terrazzo, magnesite, self-levelling, etc., also wall and floor tiling.

It is possible to have all floor coverings put under one trade described as Flooring.

Plastering

11.3	Materials	Lime Shall be high calcium lime to BS 890. Cement Shall be ordinary Portland Cement to BS 12. Water As described for 'Concrete'. Sand Shall comply with BS 1198 Type 1. Plaster Shall be retarded hemi-hydrate plaster to BS 1191. High suction plaster to be used for all blockwork applications. Scrim Shall be open weave Jute scrim. Plasterboard for ceilings To be 'Gypboard'. 9 mm thickness. Nails Shall be 30 mm long galvanized steel clout nails. Coves Shall be 'Gypcove' 75 mm deep.
11.4	Mixes—Undercoats	Brickwork: lime, cement, sand 1:1:6 mix. Blockwork: lime, cement, sand 1:1:6 mix. Plasterboard: Class B Board plaster and sand 1:3 mix. All mixes measured by volume.
	Finishes	Brickwork: neat Class B plaster. Blockwork: neat Class B plaster (see 11.00). Plasterboard: neat Class B plaster.
	Coats	Brickwork and Blockwork: 12 mm backing coat and 6 mm setting coat a.b.d. Plasterboard: 6 mm setting coat a.b.d.
11.5	Workmanship	Surfaces to be plastered to be properly set out with squared and plumbed battens with undercoats floated and ruled and scratched for key. All finishing coats to be well wetted and trowelled to a smooth and level surface from all trowel marks and other blemishes. Salient angles to be protected with metal angle beads and the plaster rounded on the arris. Junctions of dissimilar materials to be 'quirked'. Include for all timber noggings, battens, packing pieces, grounds etc., to provide level soffited and edge fixing for plaster boards.
11.6	Screeds	Screeds will be 25 mm nominal in mortar mix A as described in Brickwork trowelled smooth to receive floor finishes as later described. Lay screed a.b.d. over the G.F. slab including the garage area after applying two coats of damp-proof membrane a.b.d. to slab.
11.7	Walls	Plaster the walls, including ground floor partitions only, in two coat render and set a.b.d.
11.8	Ceilings	Fix to ground and first floor ceiling soffits plasterboard a.b.d. in full sheets where possible, scrim and set a.b.d. Fix around the perimeter of all rooms, hall, first floor landing and passages, plaster coving a.b.d. Allow for all cutting and waste.

| 11.9 | *Plaster louvres* | *Render aperture to larder vents, and set in neat plaster to both apertures fibrous plaster louvre with integral fly screen, and make good.* |

Flooring

Introduction

The vast number of floor coverings and treatments can be dealt with in specifications in different ways, and for convenience of comparison these can be divided into broad headings as follows:

Screeds—including all wet, *in-situ*, processes such as magnesite, granolithic, terrazzo, asphalt, etc. which may be continuous i.e. jointless, or divided into bays by strips of metal or plastic.
Tiles—concrete, ceramic, terrazzo, quarry, cork, thermoplastic, vinyl, rubber, linoleum, carpet, etc.
Sheets—rubber, vinyl, linoleum, etc.
Timber—blocks, boards, squares, sprung floors, etc.
Carpet—all types of carpet.
Underlay—felt, rubber, cork, etc.

On large jobs it is possible to have a *Schedule of Floor Finishes*, but generally the wet processes within the capability of the Plasterer are described under that trade. The exception to this is where specialized screeds, e.g. epoxy resin and co-polymer resins, etc. are employed and carried out by specialist firms, and described under 'Flooring' using the Prime Cost Sum procedure.

Floor tiling is invariably carried out by specialist sub-contractors, either domestic or nominated, and is dealt with under the 'Flooring' trade section.

The same remarks apply to sheet materials, but it is worth mentioning that unlike tiles, they should be left in the area in which they are to be fixed for at least twenty-four hours before they are laid, one should also remember that sheet materials, especially vinyl, may have a shrinkage factor if laid loose without adhesive.

Timber floors come in a wide range of materials and patterns, and many of the African hardwoods mentioned in the Carpentry and Joinery trade section introduction are used for this purpose, and for many years offcuts of various hardwoods, and sometimes softwoods, have been made up into squares, sometimes foil backed, as a cheaper substitute for timber blocks and boards. They are satisfactory if one accepts that there may be some uneven wear in use. Timber flooring is essentially a specialist trade and would normally be included in the 'Flooring' trade section, or in small amounts under Carpentry and Joinery. Where they are laid on ground floors it is essential to have a damp-proof membrane incorporated into the structural floor. The bitumen in which the blocks are laid is not sufficient safeguard against rising damp damage. Sealing and polishing of floors should follow carefully the manufacturer's directions.

Carpets are obviously a specialist installation, the problem for the specifier is ensuring the sub-grade upon which they are to be laid is compatible, and the advice of the carpet manufacturer should be sought before selection is made, and this will include the underlay material.

Flooring

12.0	*Carpets*	*Lay over all rooms at ground level including kitchen and cloaks, 'Finecord' carpet manufactured by Dutton Mills Ltd, P.C. £X.00 per sq. metre, laid on 'Airfoam' underlay and secured around all perimeters with edge gripper.* *Lay carpet as last described to staircase for full width of treads and risers and winders secured with angle grippers. Allow for edging strip on upper landing to meet carpet tiles a.l.d.* *Colour of carpet to selected pattern. Allow for all cutting and waste.*
12.1	*Carpet tiles*	*Lay over all rooms at first floor level, including landing and passage and both bathrooms, 450 mm × 450 mm × 6 mm 'Hugo' carpet tiles manufactured by Dutton Mills Ltd, P.C. £X.00 per sq. metre.* *Colour as selected samples.* *Allow for cutting and waste.*
12.2	*Porch area*	*Lay over the porch area 300 mm × 300 mm × 25 mm riven slate tiles as selected sample obtained from Sprowthorpe Flooring Co. Ltd, P.C. £X.00 per metre delivered to the site. Bed tiles in sand and cement mortar mix A, close, straight, jointed without grouting. Cut tiles of equal size around perimeter.* *Allow for all cutting and waste.*

Electrical works

Introduction

This section requires careful consideration in terms of the nature of the work to be done. In new works and extensions to existing buildings, and if the size of the job warrants it, tenders may be invited from specialist firms, who upon acceptance of their tender will become nominated sub-contractors in the contract documents. The Main Contractor is also entitled under the contract to tender for the work if he so wishes. In either case they will submit to the Architect a layout drawing(s) based upon the Architect's construction drawings of the proposed installation, and a specification of workmanship and materials, which must be in accordance with the current IEE Regulations and the Code of Practice. If a specialist firm is selected, the tender sum will be included in the Architect's specification, and if necessary transferred to the Bill of Quantities. See example on page 5.

In large installation works, a consultant may be employed to design the installation, and it is usual for the Consultant to undertake the inviting of tenders from the specialist firms, but the ultimate procedure in the specification will be as described. 'Package deal' firms offering design and installation services are well known in the electrical/mechanical services trades. Most of these firms are well established with excellent reputations, but discretion must be used, as the 'guaranteed performance' offered may be valid only as long as the firm remains in business.

The Architect must ensure all necessary builder's work is included, and that the Main Contractor is aware of chases and holes to be formed in the carcassing of the building so that expensive and unsightly cutting away and making good is avoided.

The Architect must also inform the tenderer of particular makes of fittings which he prefers and from whom they can be obtained.

On small works it is necessary to provide the contractor with a layout drawing and specification for the electrical work, and one can use BS 3939/1970/2, Section 27, which shows the correct symbols. British Standard Sectional List SL16 (6) is a useful guide to both installation and fittings.

Small builders will generally put the work out to a domestic sub-contractor for whom they will remain contractually responsible. The Architect will have to decide upon the type of installation to be installed in terms of the money available, e.g. if the cables are to be run on the surface or concealed, how many power points, etc. are to be provided. In housing work the ring-main system for power supply is generally adopted as it is considered to be cheaper.

In extension and conversion works one must ensure that the existing supply cable is adequate for the proposed loading, and consultation with the supply authority, and a check on the system, when proposals are drawn up may save a lot of problems later when work has begun on site. Existing systems encountered in older premises may be obsolete, or even dangerous, and may necessitate complete removal and renewal. In domestic work the demand for electricity for domestic appliances and recreational purposes has increased rapidly, and supply capacity is very important.

Electrical works

12.3	Consumer unit	Fix in the intake position shown on the plan a 'Tamel' consumer unit to BS 1454 to receive single-phase 250 V 50 Hz a.c. supply to provide the following circuits: 30 amp. ring main power circuits for ground and first floors. 30 amp. cooker circuit. 30 amp. immersion heater circuit. 15 amp. spare circuit. 5 amp. lighting circuit. 5 amp. bell circuit with transformer. The main fusebox, consumer's meter, and supply connection to be undertaken by the Eastern Electricity Board.
12.4	Wiring	All power points to be wired on the ring-main system with spurs as required in PVC cable to BS 6004 twin with earth or three core with earth. All lighting circuits to be wired on the 'looped-in' system in PVC cable to BS 6004 a.b.d. Cables will be run as far as possible in roof and floor voids properly secured and restrained with adequate clips to avoid distortion. Cables buried in walls will be protected with steel conduits to BS 31. All joints will be made at main switches, sealing boxes, socket and lighting outlets and switch boxes only.
12.5	Lighting switches	To be in accordance with BS 3676 and 4822 complete with white plastic moulded mounting boxes. Ceiling switches to be white plastic with earthing terminal.
12.6	Ceiling points	To be white plastic with connector, cover ring, rose and flex drop, terminated with a bayonet socket bulb holder in accordance with BS 67.
12.7	Socket outlets	To be 13 amp. white plastic flush pattern switched or unswitched in accordance with BS 1363. Points to be set in positions as shown on the drawings. Fixed appliances to be controlled from flush pattern white plastic spur boxes.
12.8	Heating installation	Complete in cable a.b.d. all supply circuits required for the heating installation, including all equipment as follows: 　Immersion heater 　Boiler, ignition circuit, programmer 　Pumps 　Thermostats
		Note: Night storage heaters not included in the specification. All wiring for immersion heater to be included with switch and warning light, *but heater omitted*.

12.9	Power and lighting points	Living Room	3 No. switched socket outlets

12.9 **Power and lighting points**

Living Room	3 No. switched socket outlets
	2 No. ceiling points + switch
Dining Room	1 No. socket outlet
	1 No. ceiling point + switch
Kitchen	Cooker control unit
	Pump switch
	Immersion heater socket
	outlet + switch
	Fused boiler switch
	Boiler programme
	Thermostats
	2 No. switched socket outlets
	2 No. socket outlets
	1 No. ceiling point + switch
Bedrooms 1, 2, 3	2 No. socket outlets
	1 No. ceiling point + switch
Guest room Study Nursery	2 No. socket outlets / 1 No. ceiling point + switch
Hall, landing and passages	1 No. socket outlet / 1 No. ceiling point + switch / double switching to hall, landing, and passages.
Nursery	2 No. socket outlets
	1 No. ceiling point + switch
Nursery bathroom	1 No. ceiling point + ceiling switch / water heater outlet and switch
Bathroom	1 No. ceiling point + ceiling switch
Cloaks	1 No. ceiling point + switch / water heater outlet and switch
Garage	1 No. switched socket outlet
	1 No. ceiling point + switch
Porch	1 No. Mason Rose type 7500 / flush mounted light fitment switched from the Hall.

13.0 **Water heaters**

Include the Provisional Sum of £X.00 (amount in words) for water heaters and installation charges.

13.1 **Electrical installation**

The installation shall be in accordance with the 14th edition of the Institution of Electrical Engineers Regulations for the Electrical Equipment of Buildings and the requirements of the supply authority.

13.2 **Builder's work**

The Main Contractor will include for all builder's work required in the electrical installation and making good on completion

13,3 **Intake**

Include the P.C. Sum of £X.00 (amount in words) for the service supply up to and including the intake position as shown on the drawings.
Allow for profit.
Allow for attendance.

Metalwork

Introduction

This trade section is concerned with items of purpose made metalwork, excluding those items which are normally described under other trades e.g. steel for reinforced concrete, but may include structural steelwork, particularly where this is supplied by a nominated supplier or supplied and fixed by a nominated sub-contractor.

Below are some of the items found in this trade section:

> Windows and doors.
> Rolling and sliding shutters.
> Shopfronts and grilles.
> Balustrades and handrails.
> Spiral.
> Escape staircases.

The specifier must decide where it is appropriate to include items under this section or to place them in other trades where they are associated with other materials and construction processes, e.g. it would be wrong to include in this section, wall ties, joist hangers, hip irons etc., as these clearly relate to Bricklaying, Roofing, etc., and would be described under these trades.

Metalwork

13.4	*Patio doors*	*Include the P.C. sum of £X.00 (amount in words) for the patio doors marked No. 3 on the plan to be obtained from Harvey Blake & Co. Ltd, 243 Coronation Way, Bowthorpe, Norfolk. Type 2315 delivered to site complete including glazing.* *Allow for profit.* *Fix the doors last described according to the manufacturer's instructions.* *Make good and point frame in mastic a.b.d.*
13.5	*Porch*	*Include the P.C. sum of £X.00 (amount in words) for the anodized aluminium porch unit complete with cover and supports, to be obtained from the firm last described delivered to the site complete. Type 54/22.* *Allow for profit.* *Fix the unit last described complete in accordance with the manufacturer's instructions.* *Include the provisional sum of £X.00 (amount in words) for sundries in flashings etc., as required to complete the porch.*
13.6	*Garage doors*	*Include the P.C. sum of £X.00 (amount in words) for the garage door marked No. 22 on plan to be obtained from the firm last described delivered to site complete.* *Fix the door last described complete in accordance with the manufacturer's instructions.* *Point frame a.b.d.*

Glazing

Introduction

Only the basic materials and their application are dealt with in this book. The Bibliography gives a good reference book for the history of glass making, the evolvement of modern techniques such as 'float glass', 'armourplate', etc., and the use of various types of glass in the building industry both for ordinary glazing and decorative use.

Glazing

13.7	*Glass*	*Shall comply with BS 952.* *Clear glass shall be 3·0 mm O.G.Q. sheet glass.* *Obscured glass shall be 3·5 mm narrow reeded, Group 2.*
13.8	*Putty*	*Putty shall comply with BS 544.*
13.9	*Workmanship*	*All timber rebates to be glazed shall be primed as described in Painting and Decorating.* *Allowances shall be made for expansion in cutting glass to size.* *Glass to be back puttied, sprigged, front puttied for full depth of the rebate and finished with a neat chamfer at the sight line.* *Glass secured with beading shall be set into rebates with glaziers' felt and the beads pinned in place, or fixed with cups and screws where specified, all pin heads to be punched and filled.*
14.0	*Glaze windows and doors etc.*	*Glaze windows, doors, screens all as described in the 'Window and Door Schedule' or as described in the Works.*

Painting and decorating

Introduction

The purpose of decorating is twofold, to enhance and protect the surfaces to which paint, varnish, etc., is applied. The emphasis on either of these two functions depends upon the nature of the surfaces to be decorated, and the environment in which they will exist.

Painted surfaces pose a number of problems for the specifier, particularly externally, as new materials come on to the market, claiming all sorts of advantages in their use. Whatever the claims may be, one thing is certain, the final appearance will depend upon good surface preparation coupled with the correct application of the material.

Painting is a process, albeit a skilled one, and the old argument propounded by some Contractors that 'You cannot expect a Rolls-Royce for a Mini price' is not valid. In high-class work, refinements such as stippling of the final coat of paint, are extras to be paid for. In ordinary work good painting can be achieved if proper surface preparation is carried out.

Softwood timber used in the building industry today is generally of young growth, resinous, and kiln dried. Machine planing leaves tool marks and raised grain from the outset, and if knotting and priming and subsequent coats of paint are applied to this surface without first rubbing it down smooth, these surface imperfections will be emphasized and highlighted, particularly if gloss paint is used. Nail and screw heads if not punched down or countersunk and filled, and finally rubbed down smooth, will show through the final coat of paint no matter how thickly it is applied. The removal of dust from the surfaces to be painted, the bringing forward of holes and defects with filler, and rubbing down between successive coats of paint are all essential to the quality of the finished surface.

The use of emulsion paint in lieu of proper undercoats, since they dry more quickly and are cheaper, is not unknown in the industry. The introduction of 'short oil' paints (see Painting chart in the Appendix) some years ago, and of man made resins into paints to improve drying time and achieve very high gloss paints, sometimes called 'enamels', has largely given way to the so-called 'plastic-based paints', such as epoxy resins, acrylics, polyurethanes, etc., which have numerous claims made for them. Some of these paints have proved to be incompatible with older paint surfaces to be re-painted, and they break down and flake off. They can also be affected by ultra-violet light and salty atmospheres.

The emergence of preservative/decorative coatings for timber with natural wood and coloured shades have gained ground over conventional painting.

There is no simple answer for the specifier, one finds by experience what is best to specify in given circumstances. One should observe failures in existing buildings and try to deduce why that have occurred.

Internal

Wall coverings have developed rapidly, particularly to cater for the DIY market. Tapestries, hessians, flock wallpapers, and a host of 'washables', embossed and chip faced papers, etc. Adhesives, mainly cellulose based, have become the 'norm', although starch-based pastes are still available, covering a wide range of different types and weights of paper to be hung. The Decorator will usually choose the correct paste, but it is wise to check this on site. Paper is generally pre-trimmed from the factory and clauses dealing with trimming on site can be omitted in specifications today.

Lining walls with lining paper prior to hanging wallpaper or painting is a practice which has diminished over the years, and is used mostly in older premises to improve wall and ceiling surfaces which are cracked or unkeyed. After repair of cracks and pinning back or patching of unkeyed areas, the lining paper of chosen weight can be applied in one or two layers, i.e. single or double lining, providing a good base for the paper or paint to be applied to.

Sizing of new plaster is recommended as it prevents suction by sealing the surface of the plaster; similarly, emulsion paint on new plaster surfaces is first applied as a 'mist' coat which is a thinned-out coat of emulsion paint with 10% of water added followed by two full coats to finish.

When painting asbestos, concrete, or any surface where free lime may be present, specify an alkali-resistant primer, particularly where oil-based paint is to be applied, otherwise saponification may occur which is difficult to remedy. Another problem due to soluble salts in the plaster or in the surface to which the plaster is applied is called efflorescence. Refer to Painting chart in the Appendix (page 124) for explanation of these phenomena.

External

External coatings applied to brickwork and renderings are very numerous both in composition and application, emulsion paints, chlorinated-rubber-based paints, cement-based coatings with mica and sand additives, oil and plastic-based paints, etc.

The specifier should consider carefully the environment, the compatibility with existing coatings, the time of year for application, and the general surface preparation, e.g. the use of good quality external fillers, all are essential to achieve a durable, good-looking result.

Coving and other decorative work, including renderings, is mentioned in the Plastering introduction (see page 100).

Painting and decorating

14.1	Paints	*Shall be those manufactured by Messrs Farmer & Sons Ltd, under the brand name 'Beauticote' and used strictly in accordance with the manufacturer's directions.*
14.2	*Knotting*	*Shall comply with BS 1336.*
14.3	*Fillers*	*Proprietary brands of filler may be used appropriate to internal or external work as specified.*
14.4	Primers	*For new timber to be painted shall be pink wood primer of the brand a.b.d.* *For ferrous metal shall be red oxide primer of the brand a.b.d.* *Timber specified to be stained and oiled shall receive two coats of an organic solvent teak coloured stain/preservative to comply with BS 1282 Type WB2 and finished with three full coats of teak oil.*
14.5	Finishes	*New timber:* *Knot, prime, stop, one undercoat, one finishing coat of gloss paint a.b.d. All colours to be selected and approved by the Architect and test panels painted and approved before application.*
14.6	Paint application	*Paint shall be applied only by skilled Tradesmen.* *After preparation of surfaces a.b.d. the whole to be rubbed down before undercoating and after each successive coat is thoroughly dry. The final paint surface to be free from all runs, drips, brush marks and other defects to the Architect's satisfaction.* *No external painting shall be done in wet or foggy weather conditions, and precautions taken to avoid exposure of surfaces to dampness during and after application of paint.* *All ironmongery, window and door furniture will be removed during painting and refixed when the surfaces are thoroughly dry. Oil painting on new plasterwork will not be done until the architect has given his consent. Aids to drying out, other than paraffin heaters, may be used.*
14.7	*Decorations*	*Execute all decorations as described in the Schedule of Finishes all a.b.d.*

Appendices

Door No.	F.R.	Location	Type	Lining/frame	Architrave	Ironmongery	Glazing	Hanging
1	½ hour	Front entrance	as specified	as specified	as specified	as specified	obscure	as specified
2		Kitchen (external)	FD31X26	FN26M		3 lever night latch — a/al. lever furniture	obscure	80 mm steel butts
3		Living room (patio)	as specified	as specified		delivered complete	delivered glazed	
4	½ hour	Garage to Hall	FDX26	FD26		2 lever mortice latch — a/al. lever furniture		80 mm st. r. butts
5	½ hour	Dining room	FDX29	LD29	s.w. 50 mm × 12 mm	2 lever mortice latch — a/al. lever furniture	obscure transome	8 mm st. r. butts
6	½ hour	Kitchen (internal)	FDX29	LD29	s.w. 50 mm × 12 mm	2 lever mortice latch — a/al. lever furniture	obscure transome	8 mm st. r. butts
7	½ hour	Cloaks	FDX26	LD26	s.w. 50 mm × 12 mm	2 lever mortice latch — with locking snib	obscure transome	8 mm st. r. butts
8		Staircase cupboard	FDX23	LD26	s.w. 50 mm × 12 mm	Bales catch — a/al. bow handle		80 mm steel butts
9		Living room (screen)	as specified	as specified	as specified	as specified	obscure	as specified
10	½ hour	Study	FDX29	LD26	s.w 50 mm × 12 mm	2 lever mortice latch — a/al. lever furniture	obscure transome	80 mm st. r. butts
11	½ hour	Guest room	FDX29	LD26	s.w. 50 mm × 12 mm	2 lever mortice latch — a/al. lever furniture	obscure transome	80 mm st. r. butts
12		Larder	FDX23	LD26	s.w. 50 mm × 12 mm	Bales catch — a/al. bow handle		80 mm steel butts
13	½ hour	Bathroom	FDX26	LD26	s.w. 50 mm × 12 mm	as described for door No. 7	obscure transome	80 mm steel butts
14	½ hour	Bedroom C	FDX29	LD26	s.w. 50 mm × 12 mm	2 lever mortice latch — a/al. furniture	obscure transome	80 mm steel butts
15	½ hour	Bedroom B	FDX29	LD26	s.w. 50 mm × 12 mm	2 lever mortice latch — a/al. furniture	obscure transome	80 mm steel butts
16 & 17		Wardrobes	pair LP18	FS36	s.w. 50 mm × 12 mm	Bales catch — pair 75 mm brs. bolts brs knobs		80 mm brs butts
18	½ hour	Bedroom A	FDX29	LD26	s.w. 50 mm × 12 mm	2 lever mortice latch — a/al. lever furniture	obscure transome	80 mm steel butts
19	½ hour	Nursery	FDX29	LD26	s.w. 50 mm × 12 mm	2 lever mortice latch — a/al. lever furniture	obscure transome	80 mm steel butts
20	½ hour	Nursery bathroom	FDX26	LD26	s.w. 50 mm × 12 mm	as described for door No. 7	obscure transome	80 mm steel butts
21		Linen cupboard	FDX23	LD26	s.w. 50 mm × 12 mm	Bales catch — a/al. bow handle		
22		Garage (external)	as specified			delivered complete		

Window No.	Location	Type	Size	Cills	Ironmongery	Glazing	Sundries
1	Kitchen	207C K	1200 mm × 750 mm	ext. h.w. int. s.w.	a/al. stay and fastener as supplied	O.G. clear sheet	weatherstrip
2	Cloaks	107C K	750 mm × 630 mm	ext. h.w. int. s.w.	a/al. stay and fastener as supplied	obscure	weatherstrip
3	Living room	209C K	1200 mm × 900 mm	ext. h.w. int. s.w.	a/al. stay and fastener as supplied	O.G. clear sheet	weatherstrip
4	Guest room	209C K		ext. h.w. int. s.w.	a/al. stay and fastener as supplied	O.G. clear sheet	weatherstrip
5	Study	109C K		ext. h.w. int. s.w.	a/al. stay and fastener as supplied	O.G. clear sheet	weatherstrip
6	Garage	209C K		ext. h.w. int. s.w.	a/al. stay and fastener as supplied	O.G. clear sheet	weatherstrip
7	Dining room	209C K		ext. h.w. int. s.w.	a/al. stay and fastener as supplied	O.G. clear sheet	weatherstrip
8	Larder	NL20	488 mm × 600 mm	ext. h.w. int. s.w.	a/al. stay and fastener as supplied	obscure	flyscreen — weatherstrip
9	Nursery bathroom	107C K	750 mm × 630 mm	ext. h.w. int. s.w.	a/al. stay and fastener as supplied	obscure	weatherstrip + WSM
10	Nursery	209C K	1200 mm × 900 mm	ext. h.w. int. s.w.	a/al. stay and fastener as supplied	O.G. clear sheet	weatherstrip
11	Bedroom A	209C K		ext. h.w. int. s.w.	a/al. stay and fastener as supplied	O.G. clear sheet	weatherstrip
12	Bedroom B	209C K		ext. h.w. int. s.w.	a/al. stay and fastener as supplied	O.G. clear sheet	weatherstrip
13	Bedroom C	209C K		ext. h.w. int. s.w.	a/al. stay and fastener as supplied	obscure	weatherstrip
14	Bathroom	107C K	750 mm × 630 mm	ext. h.w. int. s.w.	a/al. stay and fastener as supplied	obscure	weatherstrip

SCHEDULE OF DOORS AND WINDOWS

Location	Room space	Walls	Ceiling	Doors and frames	Architraves	Skirtings	Windows	Sundries
GROUND FLOOR	Hall & passage	sealer coat + 2 emulsion	sealer coat + 2 emulsion	K.P.S. + 2 oils	K.P.S. + 2 oils	K.P.S. + 2 oils	K.P.S. + 2 oils	
	Living room	sealer coat + 2 emulsion	sealer coat + 2 emulsion	K.P.S. + 2 oils	K.P.S. + 2 oils	K.P.S. + 2 oils	K.P.S. + 2 oils	
	Guest room	sealer coat + 2 emulsion	sealer coat + 2 emulsion	K.P.S. + 2 oils	K.P.S. + 2 oils	K.P.S. + 2 oils	K.P.S. + 2 oils	
	Study	sealer coat + 2 emulsion	sealer coat + 2 emulsion	K.P.S. + 2 oils	K.P.S. + 2 oils	K.P.S. + 2 oils	K.P.S. + 2 oils	
	Dining room	sealer coat + 2 emulsion	sealer coat + 2 emulsion	K.P.S. + 2 oils	K.P.S. + 2 oils	K.P.S. + 2 oils	K.P.S. + 2 oils	
	Kitchen	sealer coat + 2 emulsion	sealer coat + 2 emulsion	K.P.S. + 2 oils	K.P.S. + 2 oils	K.P.S. + 2 oils	K.P.S. + 2 oils	
	Cloaks	sealer coat + 2 emulsion	sealer coat + 2 emulsion	K.P.S. + 2 oils	K.P.S. + 2 oils	K.P.S. + 2 oils	K.P.S. + 2 oils	
	Entrance door & screen			sealer + 2 coats teak oil	sealer + 2 coats teak oil			
	External door to Kitchen			sealer + 2 coats teak oil				
FIRST FLOOR	Bedroom A	sealer coat + 2 emulsion	sealer coat + 2 emulsion	K.P.S. + 2 oils	K.P.S. + 2 oils	K.P.S. + 2 oils	K.P.S. + 2 oils	doors No. 17 2 coats yacht varnish
	Bedroom B	sealer coat + 2 emulsion	sealer coat + 2 emulsion	K.P.S. + 2 oils	K.P.S. + 2 oils	K.P.S. + 2 oils	K.P.S. + 2 oils	doors No. 16 2 coats yacht varnish
	Bedroom C	sealer coat + 2 emulsion	sealer coat + 2 emulsion	K.P.S. + 2 oils	K.P.S. + 2 oils	K.P.S. + 2 oils	K.P.S. + 2 oils	
	Nursery/bathroom	sealer coat + 2 emulsion	sealer coat + 2 emulsion	K.P.S. + 2 oils	K.P.S. + 2 oils	K.P.S. + 2 oils	K.P.S. + 2 oils	
	Bathroom	sealer coat + 2 emulsion	sealer coat + 2 emulsion	K.P.S. + 2 oils	K.P.S. + 2 oils	K.P.S. + 2 oils	K.P.S. + 2 oils	
EXTERNAL WORKS				sealer + 2 coats teak oil			sealer + 2 coats teak oil	
								weatherboarding 2 coats preservative to BS 1282 WB2
								fascias, barge boards, soffite boards K.P.S. + 2 oils

SCHEDULE OF FINISHES

118

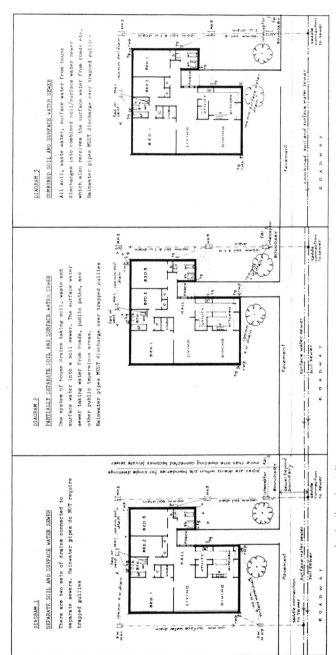

DIAGRAM 1

SEPARATE SOIL AND SURFACE WATER SEWER

There are two sets of drains connected to separate sewers. Rainwater pipes do NOT require trapped gullies

DIAGRAM 2

PARTIALLY SEPARATE SOIL AND SURFACE WATER SEWER

One system of house drains taking soil, waste and surface water into a soil sewer. The surface water sewer taking water from roads, public paths, and other public impervious areas.

Rainwater pipes MUST discharge over trapped gullies

DIAGRAM 3

COMBINED SOIL AND SURFACE WATER SEWER

All soil, waste water, surface water from house discharges into combined soil/surface water sewer which also receives the surface water from roads etc.

Rainwater pipes MUST discharge over trapped gullies

note: interceptors not required by some local authorities

Sewerage

Definition: The collection and ultimate disposal of soil and trade effluent, waste matter and surface water from drains

Means of disposal

Water carriage	Septic tank	Cesspool
Sewage plants Filtration beds Rivers and streams Sea (ebb tibed) Rivers Pollution Acts	removal of sludge and maintenance by owners	removal of sludge by local authority maintenance by owner
TYPES	**TYPES**	**TYPES**
Salt glazed stoneware BS 65 Glazed earthenware BS 540 Precast concrete BS 586	Concrete bases with brick chambers Precast concrete chambers Plastic unit complete	* Circular brick clay puddled Precast concrete

connection of drains to sewers

Joinder junctions at strategic points by local authority	purpose made junctions for insertion into sewers	saddles

Note: * this was an old method not much seen today but they do exist in some areas

Water carriage systems are of three types (see diagram in Appendix)

a. Combined system
b. Separate system
c. Partially separate

Regulations

Public Health Act 1936, Part 2
Public Health Act 1961, Part 2
Land Drainage Acts 1930–61 as amended
Rivers (Prevention of Pollution) Acts 1951–61
Greater London Council Drainage Bylaws
Building Regulations 1976, Part N

The Water Acts 1945 to date
The Water Resources Act 1945 to date cover the catchment of water, its storage and treatment.
Water Authority Bylaws with respect to installation

Drainage

Definition: The system of pipes carrying waste matter from soil, waste, and surface water pipes, to the boundary of the site, from one house or premises within the same boundary and then to the point of entry into the sewers. (Public Health Act 1875)

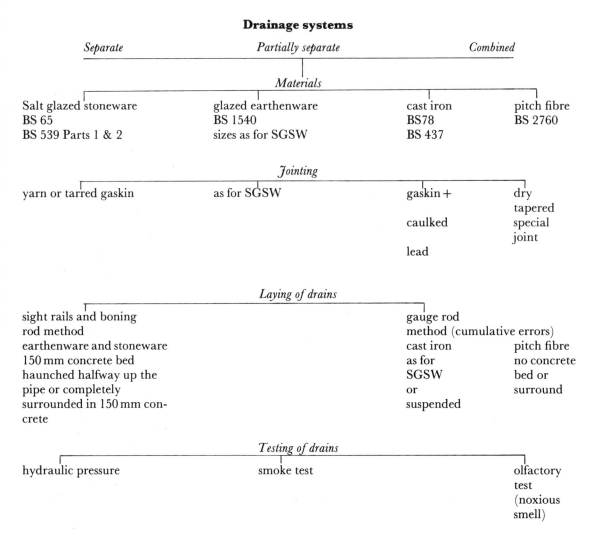

Drainage systems

Separate   *Partially separate*   *Combined*

Materials

Salt glazed stoneware	glazed earthenware	cast iron	pitch fibre
BS 65	BS 1540	BS78	BS 2760
BS 539 Parts 1 & 2	sizes as for SGSW	BS 437	

Jointing

yarn or tarred gaskin	as for SGSW	gaskin +	dry tapered
		caulked	special joint
		lead	

Laying of drains

sight rails and boning		gauge rod	
rod method		method (cumulative errors)	
earthenware and stoneware		cast iron	pitch fibre
150 mm concrete bed		as for	no concrete
haunched halfway up the		SGSW	bed or
pipe or completely		or	surround
surrounded in 150 mm con-		suspended	
crete			

Testing of drains

| hydraulic pressure | smoke test | olfactory test (noxious smell) |

Note: All new systems or extensions to existing systems must be tested to the satisfaction of the local authority (see Public Health Act 1936, Section 48. Building Regulations 1976, Part N, Public Health Act 1961, Section 16)

Other legislation to be considered:
Greater London Council drainage bylaws and Public Health (London) Act 1936, Section 34 (i).
Public Health Act 1936, Section 34 et seq.
Public Health Act 1961 Parts 1 & 5.

Plumbing

Definition: The system of pipes carrying soil, waste water from sanitary fittings, and surface water from roofs and gutters to a point of entry into the drains.

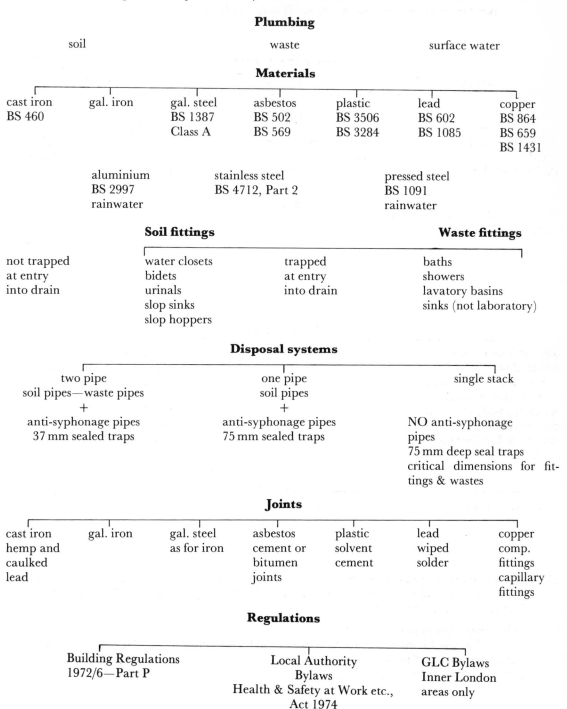

Plumbing

soil   waste   surface water

Materials

cast iron	gal. iron	gal. steel	asbestos	plastic	lead	copper
BS 460		BS 1387	BS 502	BS 3506	BS 602	BS 864
		Class A	BS 569	BS 3284	BS 1085	BS 659
						BS 1431

aluminium	stainless steel	pressed steel
BS 2997	BS 4712, Part 2	BS 1091
rainwater		rainwater

Soil fittings   **Waste fittings**

not trapped	water closets	trapped	baths
at entry	bidets	at entry	showers
into drain	urinals	into drain	lavatory basins
	slop sinks		sinks (not laboratory)
	slop hoppers		

Disposal systems

two pipe	one pipe	single stack
soil pipes—waste pipes	soil pipes	
+	+	
anti-syphonage pipes	anti-syphonage pipes	NO anti-syphonage pipes
37 mm sealed traps	75 mm sealed traps	75 mm deep seal traps
		critical dimensions for fittings & wastes

Joints

cast iron	gal. iron	gal. steel	asbestos	plastic	lead	copper
hemp and		as for iron	cement or	solvent	wiped	comp.
caulked			bitumen	cement	solder	fittings
lead			joints			capillary
						fittings

Regulations

Building Regulations	Local Authority	GLC Bylaws
1972/6—Part P	Bylaws	Inner London
	Health & Safety at Work etc.,	areas only
	Act 1974	

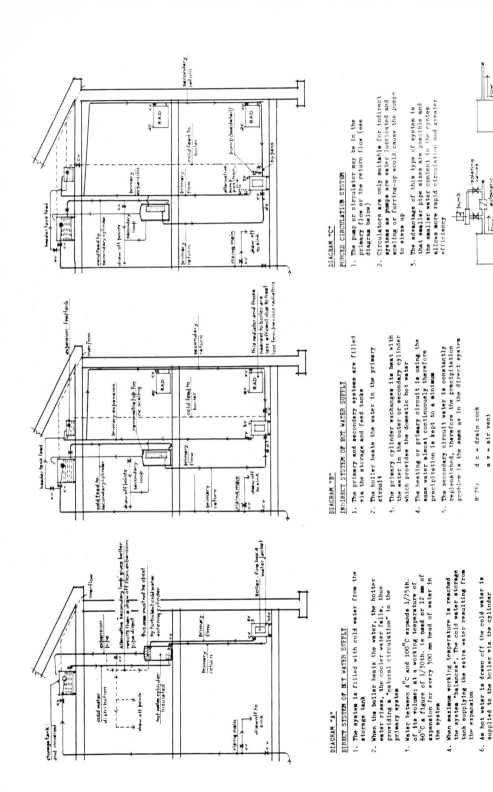

DIAGRAM "A"

DIRECT SYSTEM OF HOT WATER SUPPLY

1. The system is filled with cold water from the storage tank

2. When the boiler heats the water, the hotter water rises, the cooler water falls, thus providing a "natural circulation" in the primary system

3. Water between 4°C and 100°C expands 1/25th. of its volume; at a working temperature of 60°C a figure of 1/30th. is used or 12 mm of water in the expansion for every 300 mm head of water in the system

4. When maximum working temperature is reached the system 'balances'. The cold water storage tank supplying the extra water resulting from the expansion

6. As hot water is drawn off the cold water is supplied to the boiler via the cylinder

7. As fresh water is continually drawn into the system and heated, precipitation of carbonates, and in very hard water areas, chlorides and sulphates, will take place and eventually "fur-up" the entire system

DIAGRAM "B"

INDIRECT SYSTEM OF HOT WATER SUPPLY

1. The primary and secondary systems are filled via the storage and feed tanks

2. The boiler heats the water in the primary circuit

3. The primary cylinder exchanges its heat with the water in the outer or secondary cylinder which provides the domestic hot water

4. The heating or primary circuit is using the same water almost continuously therefore precipitation is kept to a minimum

5. The secondary circuit water is constantly replenished, therefore the precipitation problem is the same as in the direct system

N.TE: d c - drain cock

a v - air vent

c v - control valve

s v - safety valve

DIAGRAM "C"

FORCED CIRCULATION SYSTEM

1. The pump or circulator may be in the primary flow or the return flow (see diagram below)

2. Circulators are only suitable for indirect systems as pumps are water lubricated and scaling or furring-up would cause the pump to seize up

3. The advantage of this type of system is that smaller pipe sizes are possible and the smaller water content in the system allows more rapid circulation and greater efficiency

PUMP IN FLOW

PUMP IN RETURN

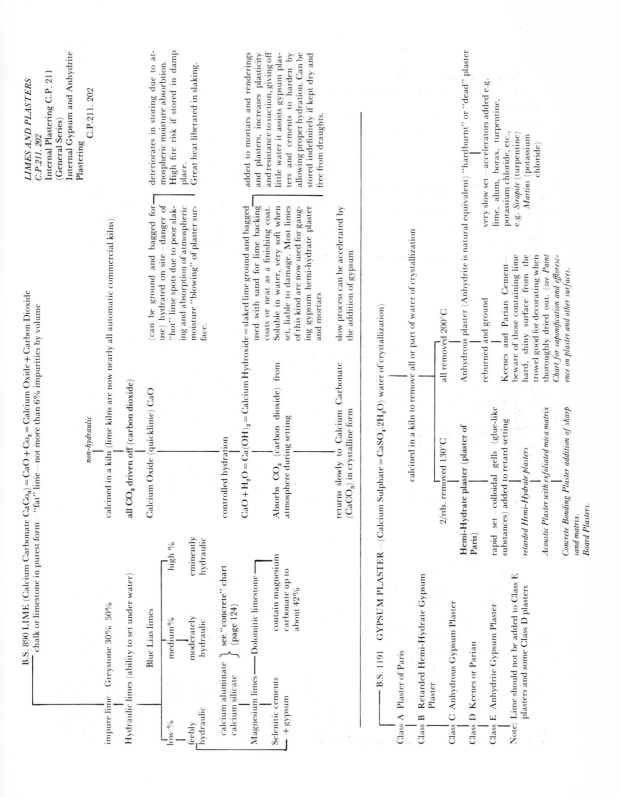

LIMES AND PLASTERS
C.P.211. 202
Internal Plastering C.P. 211
(General Series)
Internal Gypsum and Anhydrite
Plastering   C.P.211. 202

B.S. 890 LIME (Calcium Carbonate $CaCo_3$) = $CaO + Co_2$ = Calcium Oxide + Carbon Dioxide
chalk or limestone in purest form — "fat" lime — not more than 6% impurities by volume

non-hydraulic

calcined in a kiln (lime kilns are now nearly all automatic commercial kilns)

all CO_2 driven off (carbon dioxide)

Calcium Oxide (quicklime) CaO — (can be ground and bagged for use) hydrated on site — danger of "hot" lime spots due to poor slaking and absorption of atmospheric moisture "blowing" of plaster surface.

deteriorates in storing due to atmospheric moisture absorbtion.
High fire risk if stored in damp place.
Great heat liberated in slaking.

controlled hydration

$CaO + H_2O = Ca(OH)_2$ = Calcium Hydroxide = slaked lime ground and bagged used with sand for lime backing coats or neat as a finishing coat. Soluble in water, very soft when set, liable to damage. Most limes of this kind are now used for gauging gypsum hemi-hydrate plaster and mortars

added to mortars and renderings and plasters, increases plasticity and resistance to suction, giving off little water it assists gypsum plasters and cements to harden by allowing proper hydration. Can be stored indefinitely if kept dry and free from draughts.

Absorbs CO_2 (carbon dioxide) from atmosphere during setting

returns slowly to Calcium Carbonate ($CaCO_3$) in crystalline form

slow process can be accelerated by the addition of gypsum

impure lime — Greystone 30% 50%

Hydraulic limes (ability to set under water)

Blue Lias limes
high % — eminently hydraulic
medium% — moderately hydraulic
low % feebly hydraulic
calcium aluminate
calcium silicate } see "concrete" chart (page 124)

Magnesium limes — Dolomitic limestone — contain magnesium carbonate up to about 42%
Selenitic cements + gypsum

B.S. 1191   GYPSUM PLASTER — (Calcium Sulphate = $CaSO_4.2H_2O$) water of crystallization

calcined in a kiln to remove all or part of water of crystallization

all removed 200°C

2/rds removed 130°C

Anhydrous plaster (Anhydrite is natural equivalent) "hardburnt" or "dead" plaster
reburned and ground
Keenes and Parian Cement — beware of those containing lime hard, shiny surface from the trowel good for decorating when thoroughly dried out. (*see Paint Chart for saponification and efflorescence on plaster and other surfaces.*

very slow set — accelerators added e.g. lime, alum, borax, turpentine, potassium chloride, etc.,
e.g. *Sirapite* (turpentine) *Martins* (potassium chloride)

Hemi-Hydrate plaster (plaster of Paris)
rapid set — colloidal gells (glue-like substances) added to retard setting
retarded Hemi-Hydrate plasters
Acoustic Plaster with exfoliated mica matrix
Concrete Bonding Plaster addition of sharp sand matrix.
Board Plasters.

Class A Plaster of Paris
Class B Retarded Hemi-Hydrate Gypsum Plaster
Class C Anhydrous Gypsum Plaster
Class D Keenes or Parian
Class E Anhydrite Gypsum Plaster
Note: Lime should not be added to Class E plasters and some Class D plasters

PAINT TYPES

LIMEWASH (Whitewash)

fine ground lime + water
or
lime + glue size + water
(Clearcolle)

Dries by evaporation of water and slow carbonation of the lime (absorbs CO_2)

Must be removed by steam and wire brushing before painting over with oil paint – apply an alkali resistant primer first

PETRIFYING LIQUID

(petrified = stone like)

Emulsion of oil in glue solution instead of water for mixing distemper on porous surfaces.

DISTEMPER (Water Paint)

Powder pigment added to water

Not used externally used internally as a temporary decoration to assist drying out of new plaster surfaces. Must be removed before any other paint is applied (not used very much today)

OIL BOUND DISTEMPER

Oil bound water paint is pigment ground in oil (emulsion) usually a paste + water.

Exterior and interior use washable after a 3/6 months period.

More durable than water paints and may be painted over if not flaked.

(they are less used today having given way to emulsion paints)

POLISHES AND VARNISHES

FRENCH POLISH — Shellac in industrial spirit (unrefined methylated spirit)

BUTTON POLISH — Button in industrial spirit

SPIRIT VARNISH — Shellac in Alcohol

OIL VARNISH — Oil + resin approximately 2:1

FLAT OIL VARNISH — Oil + resin flatting agent e.g. Zinc Stearate + fillers e.g. Aluminium Stearate Magnesium Carbonate or Silica

FLATTING VARNISH — Hard varnish "cut down" to receive final coat

ALKYD VARNISH — A "long oil" + a synthetic resin (Alkyd)

PRIMERS –

TIMBER - SOFTWOOD — Red and White Lead in oil medium – Turpentine

TIMBER - HARDWOOD — Teak and Cedar wipe over with cellulose thinners, Oak apply one coat of aluminium wood primer

METAL - IRON AND STEEL — Remove all rust, paint one coat rust inhibitor, two coats Red Oxide Primer + two finish coats (where required)

GAL. IRON — Clean with white spirit or apply a mordant solution or Calcium Plumbate primer.

LEAD — Etching primer after rubbing down with white spirit

COPPER — Rub down with white spirit and Emery paper + any good metal primer

OIL PAINTS

consist basically of
Pigment 40/60%
Binder Oil vehicle
Thinners
Driers

Oil types — yellowing and non-yellowing

Resin — natural or synthetic and bodied oil added to basic oil paint to form a "hard gloss" paint or "enamels" i.e.

Linseed oil with boiled acid refined, or stand oil + resin % by weight

90% oil	10% resin
10% oil	90% resin

OLEO RESINOUS PAINTS

Enamels are more "brilliant" less durable mostly replaced by SYNTHETIC ENAMELS

Long oil alkyd resin varnish pigmented wholly or partly with RUTILE TITANIUM DIOXIDE

ANTIMONY OXIDE

ANATASE TITANIUM DIOXIDE

FILLERS

Stopping:

EMULSIONS

Bituminous
Chlorinated rubber
Synthetic resin types
PVA – Polyvinyl acetate
PE – Polyacrylic Esters
Polystyrene
Alkyds

Surface must be dust free, dry, free from grease. 1 sealer "mist" coat on new porous surfaces + 2 "full" coats when dry to finish.

Dries by evaporation leaving a "skin" of
Polymer
+
Pigment

Adhesion problems are now much improved. "silk" or "matt" finishes available

CEMENT PAINTS

White or coloured Portland Cements finely ground + additives

Mixed with water, applied to concrete, brickwork, rendering, etc.,

(not in low temperature) in dry conditions above 5°C to dust and grease free surfaces.

METALLIC PAINTS

Zinc, Lead, Bronze, Copper and Copper alloys stirred in a low acid medium

"flake" particles give better coverage than powders.

APPLICATION AND FAILURE OF PAINTS

The worst enemy of oil is MOISTURE alone or in combination with soluble salts, caustic acid alkalis etc., Moisture may be present in the building structure or in the atmosphere as fog, mist and rain, or by condensation conditions i.e. heat with little or no ventilation. The water must be allowed to evaporate and this means warmth and good ventilation.

SAPONIFICATION – This is a phenomenon giving rise to blistering and failure of the oil paint surface due to the chemical reaction between caustic alkali upon the vegetable oil in the paint. There is no remedy but to remove the failed paint, apply alkali resistant primer and repaint.

Beware of lime/cement backing coats to plaster – with a lime accelerator – Keenes Cement, Portland Cement, Asbestos, Stucco.

Hard filling – Whiting and Gold Size
Soft filling – Putty + Linseed Oil + Whiting
"Polyfiller" + Water external or internal quality

"Alabastine" or "Polyfiller" + Water

Concrete (C.P.114; CP115 Pre-stressed concrete)

PORTLAND CEMENT — B.S.12 Ordinary Grade

Lime = Calcium carbonate — chalk or limestone
Dicalcium silicate
Tricalcium silicate } Clay } most important
Tricalcium aluminate
Calcium aluminoferrite

+ WATER

water cement ratio =

$$\frac{\text{total weight of water in concrete}}{\text{total weight of cement}}$$

hydration of cement — evaporation of moisture related to the porosity and strength of the concrete.

AGGREGATES

coarse

fine — *critical in "designed" mixes*

Medium
B.S.882
crushed granites — fine — coarse
crushed stone   or
crushed gravel   "all-in" ballast
crushed brick — refractory
blast furnace slag
B.S.1047 (coarse only)

LIGHTWEIGHT AGGREGATES (including "no-fines" concrete)

clinker — B.S.1165 — blocks
pumice — B.S.3797 — screeds
foamed slag — B.S.877 — lightweight structural concrete
expanded clay — increased thermal insulation properties
expanded shale
pulverized fuel ash — B.S.3797 — crushing strength relative to porosity
exfoliated vermiculite
exfoliated mica
expanded perlite

HEAVY AGGREGATES

Barytes, Haematite, Linconite Iron and Steel punchings

NOTE: concrete expands 8·4 mm in 30-48 m length for 40°C change in temperature

INITIAL SETTING

"stiffening" not to be confused with "hardening" or curing reaction of concrete with tricalcium silicate — initial setting retarded in manufacture of cement by adding small amount (7%) of gypsum — Calcium Sulphate $CaSO_4 : 2H_2O$

HARDENING OR CURING

speculation — probably due to Anhydrous Calcium Silicates changing to single Calcium Silicate Hydrate

COLD WEATHER ADDITIVES

Calcium Chloride — speeds setting and hardening (about 2% by weight) always using Rapid Hardening Cement. High Alumina Cement may be used but NO additives.

AIR ENTRAINERS

increase plasticity — frost resistance, natural wood resins, animal and vegetable fats, sulphonated compounds.
Angular aggregates better gas generators — Aluminium powder 2% Hydrogen Peroxide etc.

CEMENT

Portland Cement Ordinary Grade grey-white-coloured — B.S.12
Portland Cement — Rapid Hardening and extra rapid hardening — B.S.12
Portland Cement — Sulphate resisting — B.S.4027
Portland Cement — Waterproof and Water repellent
Portland Blast Furnace Cement — B.S.146 — B.S.4246 (low heat)
Low Heat Portland Cement — B.S.1370
High Alumina Cement — B.S.915
  Ciment Fondu / Lightning Brand

INITIAL AND FINAL SETTING

reaction with water of Anhydrous Calcium Aluminate changing into Calcium Aluminate Hydrate

STRIPPING OF FORMWORK

	cold weather 2°C	*normal 16°C*
Beam sides, walls, unloaded columns	6 days	1 day
Slabs (props left under)	10 days	3 days
Beam soffits (props left under)	14 days	7 days
Removal of props beams	28 days	16 days
	35 days	16 days
slabs	21 days	7-10 days

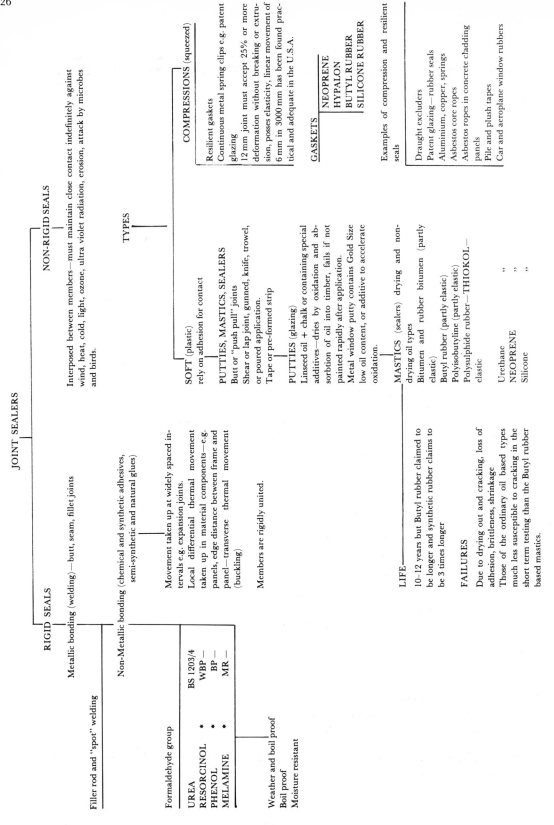

JOINT SEALERS

RIGID SEALS

Metallic bonding (welding)—butt, seam, fillet joints

Filler rod and "spot" welding

Non-Metallic bonding (chemical and synthetic adhesives, semi-synthetic and natural glues)

Members are rigidly united.

Formaldehyde group

	BS 1203/4		
UREA	WBP	—	
RESORCINOL	*	BP	—
PHENOL	*		MR —
MELAMINE	*		

Weather and boil proof
Boil proof
Moisture resistant

NON-RIGID SEALS

Interposed between members—must maintain close contact indefinitely against wind, heat, cold, light, ozone, ultra violet radiation, erosion, attack by microbes and birds.

TYPES

Movement taken up at widely spaced intervals e.g. expansion joints.
Local differential thermal movement taken up in material components—e.g. panels, edge distance between frame and panel—transverse thermal movement (buckling)

SOFT (plastic)
rely on adhesion for contact

PUTTIES, MASTICS, SEALERS
Butt or "push pull" joints
Shear or lap joint, gunned, knife, trowel, or poured application.
Tape or pre-formed strip

PUTTIES (glazing)
Linseed oil + chalk or containing special additives—dries by oxidation and absorption of oil into timber, fails if not painted rapidly after application.
Metal window putty contains Gold Size low oil content, or additive to accelerate oxidation.

MASTICS (sealers) drying and non-drying oil types
Bitumen and rubber bitumen (partly elastic)
Butyl rubber (partly elastic)
Polyisobutyline (partly elastic)
Polysulphide rubber—THIOKOL—elastic

Urethane
NEOPRENE   ,,
Silicone   ,,

LIFE
10–12 years but Butyl rubber claimed to be longer and synthetic rubber claims to be 3 times longer

FAILURES
Due to drying out and cracking, loss of adhesion, brittleness, shrinkage
Those of the ordinary oil based types much less susceptible to cracking in the short term testing than the Butyl rubber based mastics.

COMPRESSIONS (squeezed)

Resilient gaskets
Continuous metal spring clips e.g. patent glazing
12 mm joint must accept 25% or more deformation without breaking or extrusion, posses elasticity, linear movement of 6 mm in 3000 mm has been found practical and adequate in the U.S.A.

GASKETS
- NEOPRENE
- HYPALON
- BUTYL RUBBER
- SILICONE RUBBER

Examples of compression and resilient seals

Draught excluders
Patent glazing—rubber seals
Aluminium, copper, springs
Asbestos core ropes
Asbestos ropes in concrete cladding panels
Pile and plush tapes
Car and aeroplane window rubbers

Codes of Practice applicable to the Building Industry

CP 3   Code of basic design data for the design of buildings.
CP 11   Farm dairy buildings.
CP 93   The use of safety nets in constructional work.
CP 94   Demolition.
CP 95   Fire protection for electronic data processing installations.
CP 96   Access for the disabled to buildings.
CP 97   Metal scaffolding.
CP 98   Preservation treatment for timber.
CP 99   Frost precautions for water services.
CP 101   Foundations and sub-structures for non-industrial buildings less than four storeys.
CP 102   Protection of buildings against water from the ground.
CP 110   Structural use of concrete.
CP 111   Structural recommendations for load bearing walls.
CP 112   Structural use of timber.
CP 113   Arc welded construction.
CP 114   Structural use of reinforced concrete in buildings.
CP 115   Structural use of pre-stressed concrete in buildings.
CP 116   Structural use of precast concrete.
CP 117   Composite use of steel and concrete in construction.
CP 118   Structural use of aluminium.
CP 121   Walling.
CP 122   Walls and partitions of blocks and slabs.
CP 123   Dense concrete walls.
CP 131   Flues for domestic appliances burning solid fuel.
CP 142   Slating and tiling.
CP 143   Sheet roof and wall coverings—sheet metal.
CP 144   Roof coverings—asphalt and built-up roofing.
CP 145   Glazing systems.
CP 151   Doors and windows—including frames and linings.
CP 152   Glazing and fixing of glass for buildings.
CP 153   Windows and roof-lights—cleaning, safety, etc.

The above Codes of Practice can be obtained from the British Standards Institution, 101 Pentonville Road, London N1 9ND.

Asphalt derivatives—British Standards

BS 743: 1951	Material for damp-proof courses
BS 747: 1952	Classification for roofing felts (bitumen and fluxed pitch)
BS 802: 1955	Tarmacadam with crushed slate or rock aggregate
BS 988: 1957	Mastic asphalt for roofing
BS 1070:1956	Black paint – tar based
BS 1076:1956	Mastic asphalt for flooring
BS 1097:1958	Mastic asphalt for tanking and damp-proof courses
BS 1262:1957	Mastic asphalt for roofing (natural rock asphalt aggregate)
BS 1241:1959	Tarmacadam and tar carpets (gravel aggregates)
BS 1242:1945	Tarmacadam for paving
BS 1310:1950	Coal for tar pitches for building purposes
BS 1324:1946	Asphalt tiles for flooring or paving (natural rock asphalt)
BS 1375:1947	Coloured pitch mastic asphalt
BS 1410:1959	Mastic asphalt for flooring
BS 1418:1954	Mastic asphalt for tanking and damp-proof courses (natural rock asphalt)
BS 1450:1948	Black pitch mastic flooring
BS 1451:1956	Coloured mastic asphalt flooring (limestone aggregate)
BS 1621:1954	Bitumen macadam with gravel aggregate
BS 1690:1950	Fine cold asphalt
BS 1783:1951	Coloured pitch mastic flooring incorporating lake asphalt and bitumen
BS 2040:1955	Bitumen macadam with gravel aggregate
BS 3051:1959	Coal tar oil types of wood preservative

List of abbreviations

The following list of abbreviations used in the building industry may be used to draft specifications, Bills of Quantity, and other documents.

Lists of terms and abbreviations will also be found in the British Standards – Glossaries of Terms covering a wide range of processes, components, and materials in the building industry. The information should be sought in the British Standards Sectional List SL16 to identify the particular Standard you require, e.g. BS 5578: Part 1: 1978 – Building Construction – Stairs – Vocabulary, which lists the terms used in staircase construction.

The use of proper terms in relation to building matters indicates a good professional training and standard. There are differences in terminology both inthe United Kingdom and abroad, and these have to be learned when the need arises.

List of abbreviations used in specification writing and in quantity surveying

A.B.	Air brick	clg	ceiling
a.b.	as before	c.o.	consisting of
a.b.d.	as before described	comp	complete
a.d.	as described	comp'n	completion
a.l.d.	as last described	c on p	circular on plan
adj	adjust(ment)	C.P.	chromium plated
agg	aggregate	c & p	cut and pin
alt	alternate	cpg	coping
appvd	approved	cpr	copper
Archt	Architect	crs	course
archive	architrave	c.s.a.	cross sectional area
ard } aro }	around	C.S.G.	clear sheet glass
		csg	casing
asph	asphalt	csnk	countersunk
art	artificial	ct.	cement or coat
av } ave }	average	ctd	coated
		ctg	cutting
		cupd	cupboard
B	brick	C.W.	cold water
BD	bill direct	c & w	cutting and waste
bd	board	C.W.S.T.	cold water storage tank
bd d	boarded		
bdg	boarding	ddt	deduct
b i	build in	delvd	delivered
bit	bitumen	dep	deposit
BJ	black japanned	dia	diameter
b.j.	breaking joint	dist	distemper
b & j	bed and joint	div	divided
b & p	bed and point	dp	deep
bk	brick	d.p.c.	damp-proof course
bkwk	brickwork	dr	door or drain
bldg	building		
BMA	bronze metal antique	ea	each
brkts	brackets	el	extra large
brrs	bearers	e.l.p.	extra large pipe
brs	brass	e.m.l.	expanded metal lath
B.S.	British Standard	emuls	emulsion (paint)
b.s.	both sides	eng	engineering
b.s.m.	both sides measured	E.O.	extra over
bsmt	basement	e'ware	earthenware
btm	bottom	ex	exceeding
bwk	brickwork	exc	excavate
		exc'n	excavation
c.a.	cart away	exp	exposed
cantr	cantilever	extg	existing
cav	cavity	ext'ly	externally
c.b.	common brickwork		
c.c.n.	close copper nailing	F.A.I.	fresh air inlet
ccs	centres	F.A.O.	fresh air outlet
c & f	cut and fit	fast	fastener
chal	channel	fcgs	facings
c.i.	cast iron	f f	fair faced
circ.	circular	fin	finished
c.l.	centre line	fin'gs	finishings

f.l. & b.	framed, ledged and braced		L.A.	local authority
flem	flemish		lab	labour
flr	floor		L.B.	lavatory basin
fmwk	formwork		len	length
fndn	foundation		levs	levels
F.O.	fix only		l & h	lime and hair
fr	frame		lin	linear
fsnr	fastener		l & m	labour and materials
furn	furniture		l.o.	linseed oil
fwd	forward		l.p.	large pipe
fxg	fixing		l & p	lath and plaster
fxd	fixed		l & r	level and ram
g	girth		m	metre
galv	galvanized		mat'l	material
g.b.d.p.	galvanized barrel distance piece		max	maximum
gn'l	general		meas sep	measured separately
g.i.	galvanized iron		med	medium
g.l.	ground level		met	metal
glzg	glazing		m.g.	make good
grnds	grounds		m.h.	manhole
grve	groove		mi	mitre
g.s.	general surfaces		min	minimum
			mis	mitres
H.B.	half brick		mld	moulded
	(for free standing walls)		mm	millimetre(s)
	$\frac{1}{2}$B for additional wall thickness		m.m.s.f.	machine made sand faced
h.b.s.	herring bone strutting		mo	moulded
h'core	hardcore		m.s.	measured separately
hd'bd	hardboard		m.s. or MS	mild steel
hdg	heading		msd	measured
hi	high		mull	mullion
h n & w	head nut and washer			
h.m.s.f.	hand made sand faced		n/e	not exceeding
horiz	horizontal		necess	necessary
h.p.	high pressure		No.	number
H.R.	half round		nom	nominal
h'rl	handrail		n.w.	narrow widths
hsd	housed			
H.W.	hot water		o/a	overall
h.w.	hollow wall		o.c.n.	open copper nailing
hw	hardwood		o'hang	overhang
			o'flow	overflow
inc	including		o.s.	oversite
int'ly	internally		o'slg	oversailing
irreg	irregular		o.s.o.	one side only
inv	invert (level)		O.T.	open top
jnt	joint		pat'n	pattern
jst	joist		P.B.	polished brass
junc	junction		p.b.	plasterboard
			P.C.	prime cost
kg	kilogramme		pcé	piece
km	kilometre		P.ct.	Portland cement
K.P.S.	knot, prime & stop		ped	pedestal
			plas	plastic
l	length or long			

plstr	plaster		soff	soffit
p.m.	purpose made		s.p.	small pipe
pnl	panel		spcg	spacing
pntg	pointing		splyd	splayed
p.o.	planted on		sq	square
P.O.	prime only		s & s	spigot and socket
pos'n	position			sides and soffits
ppt	parapet		surf	surface
pr	pair		surr	surround(s)(ded)
prep	prepare(d)		susp	suspended
proj	project(ion)		S & VP	soil and vent pipe
prov	provisional		s.w.	soft wood or stoneware
P.S.	provisional sum		S.W.G.	standard wire gauge
p.s.	pressed steel			
p & s	planking and strutting		T.C.	terra cotta
p & scr	plug(ged) and screw(ed)		t & g	tongued and grooved
ptn	partition		th	thick
			thro'	through
q.t.	quarry tile		thrtd	throated
			tngd	tongued
rad	radius		tog	together
R.C.	reinforced concrete		tr	trowelled or trench
reb	rebated		t & r	tread and riser
rec	receive			
red	reduce(d)		u/c	undercoat
reinf	reinforce(d) (ment)		u/s	underside
rem	remove			
ret'd	returned		veg	vegetable
ret'n	return		vert	vertical
r f & r	return, fill and ram		vit	vitreous
r.h. & s.	rivet head and snap		V.P.	vent pipe
rkg	raking			
r.l.	reduced level		W.C.	water closet
rl	rail		wdw	window
rnd'd	rounded		W.G.	white glazed
ro	round or rough		wh	wheel
r.o.j.	rake out joints		wi	width
r.f. & s.	render, float and set		w.i.	wrought iron
R.W.P.	rainwater pipe		wk	work
			W.P.E.	white porcelain enamel
S.A.A.	Silver anodized aluminium		wrot	wrought
san ftgs	sanitary fittings		W.W.P.	water waste preventer
s.c.	stop cock			
sckt	socket		x-gr	cross grain
s.d.	screw down (valve)		x-reb	cross rebated
scr	screw		x-tngd	cross tongued
scr'g	screwing			
s.j.	soldered joint		Y.S.	York stone
sktg	skirting			
s.l.	short length		Sundries:	
s & l	spread and level(led)		Ø	diameter
sm	smooth		>	angle
snk	sunk		② ③	2 or 3 coats of oil paint
			2'ce	twice

Bibliography

British Standards Handbook No. 3, British Standards Institution.
Building Materials and Components for Housing, British Standards Institution.
Mitchell, *Building Construction (Advanced Course)*, 13th edition, Batsford.
King and Everett, *Components and Finishes*, Batsford.
Persson, C., *Flat Glass Technology*, Butterworths.
Worthington, J., *Home Electrics*, Orbis Publishing.
Introduction to Concrete for Students, Cement and Concrete Association.
Everett, A., *Materials*, Batsford.
Slat Glazed Drainage Material, Doulton Vitrified Pipes Ltd.
Sectional List of British Standards, SL 16, British Standards Institutional.
The Building Regulations 1976, HMSO.
The London Building Acts (Amendment) Act 1939, GLC Publications.
The JCT Guide 1980, RIBA Publications Ltd.
The JCT Standard Form of Building Contract 1980, RIBA Publications Ltd.

Index

Excavation

1.0	Site	Trial holes indicate vegetable soil to a depth of approximately 300 mm, the subsoil is medium clay to an average depth of 2400 mm.
1.1	Definition	The term excavate shall mean digging, getting out spoil of any material encountered, loading into barrows, wheeling, spreading and levelling on site as directed, or removing to tip as required.
1.2	Planking & strutting	Supply and fix all timber planking and strutting to support excavations until backfilled, strike and remove as the backfilling proceeds.
1.3	Removal of water	Keep all excavations free from water by pumping or baling as required, and dispose of same in such a way as to avoid nuisance to adjoining property.
1.4	Hardcore	Shall be clean, dry, broken brick or stone, free from all deleterious matter and graded from 150–50 mm sieve. Lay in layers not exceeding 225 mm deep at any one time, ram and consolidate to required reduced levels.
1.5	Re-use of old materials	Existing materials, brick, stone, concrete, from the demolition works may be used as hardcore if complying with Clause 1.4 a.l.d
1.6	Blinding	Blind hardcore after consolidation with a layer of clean, dry, ashes, or weak cement slurry. Note: plastic sheeting and building paper may be used for blinding
1.7	Strip site	Excavate over the area of the building and for a distance of 1500 mm beyond the perimeter walls, to a depth of 300 mm to remove all vegetable soil and deposit on the south boundary of the site in heaps not exceeding 1 m high. Note: or remove to tip as required. Existing turf would be cut and stacked for sale or disposal as directed by the building owner
1.8	Foundation trenches (see page 28)	Excavate the foundations trenches to the widths and depths as shown on the drawings, retain sufficient selected spoil for backfilling a.b.d. and remove remainder to tip. Ram and consolidate trench bottom.
1.9	Hardcore beds (see page 28)	Lay hardcore as shown on the drawings to the thicknesses and levels as required, consolidate and blind a.b.d.
2.0	Drain runs	Excavate drain trenches, manholes, to the required widths, lines, and gradients as shown on the drawings, backfill and consolidate on completion of drainlaying a.b.d.

Drainlaying

Introduction

The charts in the Appendix (see pages 119/120) give a synoptic view of both sewerage and drainage, and the systems and materials in general use. The responsibility for both systems rests with the local authority, whose terms of reference are the Public Health Act 1936, and the Building Regulations, Parts N & P.

In the Inner London Boroughs of the Greater London Council the Public Health (London) Act 1936 obtains, together with the London Building Acts (Amendment) Act 1939, and the Local Government Act 1939, Section 147. The Greater London Council publishes a Code of Practice for drainage in its areas.

The provisions of the Health and Safety at Work etc., Act 1974, Part 3, applies to all areas other than those under the jurisdiction of the Greater London Council in the Inner London Boroughs and the City of London, although Section 70 of the Act makes provision for overall control by Central Government.

Septic tanks and cesspools are permitted where a water carriage system of sewerage does not exist, and both standard drawings and specifications are available from local authorities, together with a number of proprietary systems available on the market.

The work on site will be inspected by the Environmental Health Officer of the local authority, who will check the system installed against the drawings and standard application form submitted for approval, and will also supervise testing of the system on completion.

The drainage system, and this may also include plumbing works, may be undertaken by specialist consultants and contractors, and the P. C. sum procedure would obtain.

On large jobs a manhole schedule may be drawn up with numbers of manholes corresponding to those shown on the drawings. Double-seal manhole covers are required internally, otherwise single-seal covers are used. Where vehicular traffic is to pass over manholes heavy-duty covers must be specified.

Road gullies adjacent to areas where silt will be washed down with surface water into drains should include for a silt tray to facilitate easy silt removal.

Drainlaying

2.1	Cement	*As described for Concrete.*
2.2	Sand	*As described for Concrete.*
2.3	Water	*As described for Concrete.*
2.4	Concrete bedding	*Shall be 'all-in' ballast concrete mix A as described in Concrete. Bedding shall be 150 mm deep, twice the dia. of the pipe wide, and haunched half way up the dia. of the pipe. Hand holes to be left for jointing and making good after pipes have been tested.*
2.5	Pipes	*Soil pipes shall be salt glazed stoneware to BS 65 British Standard Tested Quality.* *Surface water pipes shall be salt glazed stoneware to BS 65 British Standard Quality.*
2.6	Gullies	*Shall be salt glazed earthenware to BS 539.*
2.7	Jointing	*Salt glazed ware pipes to be jointed with tarred gaskin and sand and cement 1:1 mix used freshly mixed and well rammed into joint and finished with a neat chamfer around the socket. All pipes to be drawn through to remove all debris as the work proceeds.*
2.8	Pipes passing through walls	*Pipes passing through walls shall have a minimum clearance around the pipe of 25 mm and be wrapped in bitumen-impregnated felt.*
2.9	Manholes (inspection chambers)	*Base to be 150 mm thick concrete mix B as described in Concrete with 150 mm projection beyond the external chamber dimension. Form the chamber in semi-engineering bricks a.b.d. 225 mm thick in English bond in mortar mix A a.b.d. in Brickwork. Corbel chamber to receive frame and cover as later described.*

Main channels and branch drains

* If bends are sharp specify $\frac{3}{4}$ round channels, if slow specify $\frac{1}{2}$ round channels. State the angle if accurate or leave to the builder to set out. A drain chute may be specified if the amount and velocity of the flow warrants it, and if rodding of the drain may be difficult. Include for an interceptor if required in the system.
Branch drains to discharge 50 mm above the main channel.

Bench up main channel and branches in mix A mortar a.b.d. trowelled smooth to outfall.

Note: rendering of manhole chambers is a practice that is not recommended any more.
If the condition of the sub-soil warrants it a sulphate-resistant cement should be specified for the manhole mortar.

	*	Step irons Where chamber is more than 1 m deep to be galvanized hoop irons set in every fourth brick course staggered at 300 *mm c/s.*
	*	Fresh air inlet To be Messrs Famptons type FAI/12 complete with brass frame and mica flap. Position and jointing to system to be shown on drawings.
		Manhole covers shall be to BS 497 Grade B—Table 6. 610 mm × 406 mm clear opening size. *Set frame in sand and cement mortar.* *Mix A a.b.d. set covers in red lead putty.*
3.0	Lay drains	*Lay all drains to the lines and gradients as shown on the drawings, bedded, haunched, and jointed all a.b.d. to include all manhole, interceptor, gully, and W.C. connections as shown. Backfill after testing as described in Excavation.*
3.1	Testing of drains	*On completion of laying of drains the whole system will be tested to the satisfaction of the local authority and the architect, and again on completion of the contract, any part of the system found to be faulty will be taken up and replaced until the system is passed as satisfactory, such remedial work to be at the contractor's expense.* Testing shall be as laid down in BSCP 301.
3.2	Septic tank	*Fix in the position shown on the drainage a 'Leps' septic tank manufactured by Spelman, Harris & Co. Ltd, Harrogate, Yorks. Install complete in accordance with the manufacturer's instructions, and connect to drainage system a.b.d. Final effluent to be discharged into the existing watercourse as shown through land drains to BS 1196.*
	*	Where a connection to the public sewer is to be made in a water carriage main drainage system, it will be dealt with as a P.C. Sum for the local authority to carry out the work.
	*	Soakaways would normally be described in Excavation work in positions as shown on the drawings, and the size would vary according to the amount of surface water to be disposed of. An average size would be 2000 mm deep × 2000 mm dia. the bottom consolidated and filled with graded rubble and hardcore with land drains to disperse the water over a large area or take it to a watercourse. In the former case the drains would be perforated.

Concrete

Introduction

The range of concrete work makes the trade one of the longest to be specified, from mixing a small amount of concrete by hand to highly sophisticated structures with controlled mix designs, various forms of reinforcement, including pre-tensioned and post-tensioned work, designed formwork, and a host of other refinements both *in situ* and precast. The Codes of Practice and British Standards covering all concrete processes are very numerous, providing much useful information on the materials used, the site and workshop processes, site and laboratory work, including testing etc.

Large organizations such as The Cement and Concrete Association Ltd, offer a wide range of literature free or at nominal charge, which can be invaluable to the specification writer. The following points, among others, will arise in specification writing in addition to that described in the trade section:

1 Short-bored piling	Reinforcement and mix (see notes in Excavation introduction).
2 Reinforced foundations	Reinforcement and mix – bars to BS 785 – fabrics to BS 1221 – bending to BS 4483 and BSCP 114.
3 Column and stanchion casing and steel beam casing	Reinforcement and mix – splays to other labours – surface treatments – cover thickness – binding wire.
4 Staircase *in situ* or precast	Reinforcement and mix *in situ*, refer to drawings for dimensions, thicknesses, surface treatments. Ditto precast or P. C. Sum for specialist supplier, allow for builder's work for building in and making good.
5 Underpinning	Extent of work as specified or as shown on the drawings (shoring, etc. taken in Excavation, see introduction to this trade page 66).
6 Kerbs	BS 430 – size, type, bedding.
7 Paving	Paving slabs to BS 368 – size and thickness, pattern of laying, jointing, bedding, pointing.
8 Concrete pavement lights	P.C. sum for specialist suppliers, allow for fixing if not included in P.C. sum
9 Hollow tile slabs	*In situ* work or specialist construction.
Precast floors	*In situ* – reinforcement and mix, formwork, dimensions as drawings.
Roof slabs	Hollow clay blocks to BS 3921, end and side bearing, beam sizes, slip tiles, holes to be formed.
	Precast work – Specialist work on P.C. sum, supply and fix or supply only. Allow for building in and making good.

10 Generally   Ties for blockwork and brickwork to concrete to BS 1243,
                                                 allow for timber plugs, metal and plastic fixings, etc.

In large works the whole of the concrete work may be undertaken under the supervision and control of a structural engineer consultant, who will provide all the design calculations, drawings, and specification, and will supervise the work on site, which may be in the role of Site Engineer.

The 'package deal' organizations who design and erect concrete structures on the basis of an estimate has been discussed in Chapter 1. Items such as precast portal frames, system type buildings, storey height and other forms of cladding panels are very prevalent in the building industry, and the choice for the specifier is one of supply only or supply and erect. The attendant items such as glazing, pre-surface treatment of the structure, and jointing techniques, must be considered carefully and covered in the specification. Protection of the cladding after erection is also necessary. Check that the fixing systems have approval of the local authority building control officers, and if an Agrément Certificate exists for the system.

Surface treatments such as bush hammering, exposed aggregate, board marked surfaces, profiles etc., are specialist treatments requiring technical knowledge before they can be specified properly. The reference material referred to in this introduction will be helpful in this regard. Water stops, expansion joints, bituminous and other surface applications are also dealt with.

There is some confusion over terms such as precast concrete, reconstructed stone, and artificial stone. Artificial stones based upon Howe's classification: (*Building Construction*, Mitchell, 13th edn, Batsford)

1 Stones made of natural rock fragments held together with cement such as Portland Cement.
2 Stones made as described in 1 above, but treated with a silicate of soda process (placed in tanks for prescribed periods then taken out and cured).
3 Stones in which rock has been pulverized or sand cemented with carbonate of lime.
4 Stones in which more or less carbonate of lime is replaced with silicate of lime.
5 Stones cemented by bituminous, asphaltic, or other organic substance.

Indurated concrete or artificial stone made from pulverized granite or sandstone cemented with Portland Cement treated with silicate of soda as described in 2 above.

Reconstructed stone made under Thom's patent is usually the debris of limestone quarries ground to dust, mixed with Dolomite lime, and then run with water and autoclaved under great pressure and heat. The blocks are then dried, placed in a vacuum, carbon dioxide is admitted and carbonization takes place to completion. Sedimentary rocks can be reconstructed by this process.

Concrete

3.3	*Cement*	Shall be Portland Cement ordinary grade to BS 12.
3.4	*Storage*	All cement shall be sorted under cover on raised platforms, kept clean and dry, and used strictly in delivery sequence.
3.5	*Aggregates*	a. Aggregates for mass concrete shall be Thames Ballast to BS 882. b. Fine aggregate for all concrete shall be clean, sharp, pit or river sand to BS 882. c. Coarse aggregate shall be crushed gravel graded to 18 mm to BS 882. d. 'all-in' aggregate shall be fine and coarse aggregate ready mixed to BS 882.
3.6	*Water*	Shall be clean (drinkable), free from all organic impurities or deleterious matter and to be at a temperature of 5°C and rising. Only sufficient water to obtain a workable mix, or as gauged by water/cement ratio scale, shall be used.
3.7	*Reinforcement*	Shall be round section mild steel bars to BS 4449 free from all loose mill scale, rust, and grease, and thoroughly wire brushed before placing in the formwork. Reinforcement will be cut and bent as described in the bending schedules or as specified, and placed in accordance with the drawings or as specified, and tied with galvanized soft iron wire.
3.8	*Spacing*	The required concrete cover to reinforcement shall be maintained by the use of suitable spacers in the formwork in accordance with BSCP 114.
3.9	*Formwork*	All formwork shall be sound, well-seasoned softwood, clean on all surfaces, properly secured to retain concrete without any distortion or movement during setting, and struck in accordance with BSCP 114 recommendations. Concrete specified as 'fair faced' shall be placed in formwork having a grade B/B plywood lining rubbed down smooth and well oiled. All formwork so constructed shall produce the exact profiles as shown on the drawings.
4.0	*Measuring*	Materials for mass concrete shall be measured by volume in batch boxes of proper size, thoroughly cleaned out at the end of each working period. Materials for reinforced concrete shall be measured by weight in a batch mixer with an accurately calibrated scale.
4.1	*Mixing*	Only sufficient water as described in Clause 3.6 shall be used to obtain a workable mix of uniform colour. In hand mixing increase the cement content by 5%.
4.2	*Placing*	All concrete to be placed in the forms immediately after mixing. No concrete shall be reconstituted or re-used after a period of one hour after mixing. No concrete shall be placed in forms to a

greater height than 1 m at any one time. Mass concrete to be tamped into excavations and forms to prevent voids.

Reinforced concrete to be hand or shutter vibrated in accordance with BSCP 114 recommendations or for such periods as may be specified, to ensure a dense mix free from honeycombing and voids.

4.3	Curing	*After placing in the formwork concrete shall be cured by hosing at intervals, or by application of wet sacking for two days after placing, or by use of PVC or polythene sheeting methods as BSCP 114 recommendations.*

4.4	Temperature	*No concrete shall be placed in temperatures below 2°C except with the consent of the Architect or Engineer.*

4.5   Mixes

Type	cement	'all-in' agg.	fine agg.	coarse agg.
A	1	8		
B	1	6		
C	1		2	4
D	1		$1\frac{1}{2}$	5

4.6	Foundations	*Form strip foundations to the widths and depths as shown on the drawings in concrete mix B a.b.d. level to receive bkwk.*

4.7	Slabs	*Lay, spread, and level 150 mm thick ground floor slab to include porch entrance and step to kitchen as shown on the drawings in concrete mix B a.b.d.*

4.8	Damp-proof membrane	*Lay over the slab last described two coats of bituminous membrane a.b.d. protect until screed is laid.*

4.9	Lintols	*Form all lintels to the sizes specified and as shown on the drawings, reinforce as specified, perform all labours in forming drips, rounded arrises, hacking for key etc., as required.* *Perform all labours in getting into position, bedding and making good. In-situ lintols to be cast in concrete mix C a.b.d.*

Brickwork and blockwork

Introduction

The choice for the specifier of brick types, sizes, jointing, and mortar mixes is a wide one. In specifying brickwork the choice is determined usually by the amount of money available for the bricks in relation to the total cost of the works. It may also be conditioned by the locality of the work to be done, e.g. the planning authority in the area may require a particular type of brick to harmonize with surrounding buildings.

Brick manufacturers and suppliers will often quote for bricks 'ex works', i.e. from the brickworks, and if the choice is a hand-made Leicester facing brick for a job in Cornwall, there will be a high transport cost over and above the 'ex works' price quoted. Hence, the following example of brick specification underlines the phrase 'delivered to the site' to remind you of this fact.

Local brickworks still exist in some areas producing annually a limited number of bricks from their kilns, and these will be characteristic of the local clay or brick earth in colour and texture.

Be careful with bricks which are termed 'sand faced', which are made from brick earth of the Fletton kind, with an applied facing of sand of various colours on one header and one stretcher face. In exposed conditions over a long period of time the sand face may disappear and leave a Fletton brick appearance for which there is no remedy.

Soft bricks and mortars should not be used under ground as they will be subject to frost action and deteriorate. Sulphate-resistant cements should be used as a precaution against sulphate attack from the soil.

Among a number of publications in the clay industry the *Brick Bulletin* and the *Building Research Station Digests* offer useful information on bricks and their use.

Exercise great care in specifying work to old buildings, particularly where walls appear to be very thick, and were built in lime mortar. Such walls may prove to be no more than brick skins filled with rubble and bound together with a lime slurry, with sometimes timber 'bonders' inserted to spread the load in the wall. In removing brickwork to form openings in such walls it is unwise to use any form of mechanical tool. Bricks should be carefully removed by hand with the minimum of hammering.

The author once specified an opening to be formed in a period house basement for a slow combustion stove, prohibiting in the specification the use of mechanical tools. The builder failed to comply with the specification requirement, and the use of a mechanical hammer caused the flue withes over eight floors to collapse into the basement and ground floor chimney breasts. The builder had to make good the damage at his own expense at great cost. Without the prohibiting clause the position would have been very different for the specifier.

Do not specify strong mortars for re-pointing old brickwork, specify a gauged mortar compatible with the density of the bricks to avoid shrinkage and frost damage. See Brickwork trade for summer and winter mortar mixes.

Mortars change the appearance of brickwork, and sample bricks delivered to the office with white felt simulated joints may look very different when built and bonded with a particular mortar mix and joint. A sample panel or two built on site with varying joints and mortar mixes is not an expensive thing to do and may save disappointment with the face brickwork.

Brickwork and blockwork

5.0	Bricks below d.p.c.	To comply in all respects with BS 392, Part 2, and to be semi-engineering quality P.C. £X.00 per thousand delivered to the site.
5.1	Bricks above d.p.c.	Common bricks to be 'Fourgroove' or other grooved Flettons of equal quality. P.C. £X.00 per thousand delivered to the site.
5.2	Facing bricks	To be Midland hand-made facing bricks manufactured by The Midland Brick Co, Ltd, P.C. £X.00 per thousand delivered to the site.
5.3	Unloading and stacking	The Contractor will unload, get into the site, store and protect all bricks delivered to the site.
5.4	Protection of face brickwork	All facing brickwork to be protected with polythene sheeting secured with timber battens immediately after erection of the brickwork until completion of the contract.
5.5	Cement	Shall be Portland Masonry Cement above d.p.c. level. Sulphate-resisting cement to BS 4027 to be used below d.p.c. level.
5.6	Sand	Sand for mortar shall comply in all respects with BS 1200.
5.7	Lime	Shall be high calcium lime to BS 890.
5.8	Plasticizers\nNote: this agent is incorporated in Masonry Cement	Plasticizers may be used at the discretion of the architect and used strictly in accordance with the manufacturer's directions.
5.9	Mortar mixes	Clay bricks and concrete blocks

Mix	lime	cement	sand
A	—	1	3
B	2	1	8–9
C	1	1	5–6

Mix B to be used in Spring and Summer
Mix C to be used in Winter

6.0	Partition blocks	Shall be 'Porocrete' lightweight cellular blocks 75 mm or 50 mm thickness as shown on the drawings or as described in the specification. Lay blocks in mortar mix B and block bond at abutments.
6.1	Wall ties	Shall be heavy pattern galvanized steel to BS 1243.
6.2	Damp-proof course	Shall be 'Bitumex' bitumen-impregnated fibre reinforced hessian manufactured by Bitumex Products Ltd, laid full width and length of walls, stepped as required, with laps in the length not less than 100 mm as necessary. Trim to the face of the brickwork and point a.b.d.
6.3	Flashings	Under cills lay 'Bitumex No. 1' flashing material 75 mm longer

than the cill at each end and for the full width of the wall. Trim and point a.b.d.

Stepped flashings in cavity walls over window and door openings shall be 22 SWG copper 150 mm longer than the opening at each end, 150 mm deep and turned for full width into the inner and outer cavity wall skins.

Note: Other flashings described under Plumbing and Roofing as appropriate.

6.4	*Cavity walls*	*Below d.p.c. the inner and outer skins to be constructed in bricks and mortar as described for work below d.p.c. level, and the cavity filled with concrete mix C to one course below d.p.c. level. The two skins to be tied together with ties a.b.d. not less than 2 ties per square metre. Cavity walls above d.p.c. to be built in two skins, the outer skin up to weatherboarding level in facing bricks a.b.d., the inner skin and outer skin above weatherboard level in concrete cellular blocks laid in mortar mix B a.b.d., the two skins tied together with ties a.b.d. 450 mm vertically and 1000 mm horizontally staggered, with additional ties at 225 mm c/s around openings. Close the cavity at openings against the outer skin with blocks a.b.d. and a damp-proof course of 'Bitumex' a.b.d. set in mortar for full height of the reveals.* *Leave adequate weepholes over heads of openings in cavity walls, and keep the cavity clear of all debris and mortar droppings during construction.* *Form all window and door openings in the positions and to the sizes as shown on the drawings, build in lintels, door frames and windows as specified.*
6.5	*Pointing*	*Facing brickwork to be finished with an ironed joint as work proceeds.*
6.6	*Generally*	*Brickwork to be laid level and plumb and all perpends vertical with the bond, bats will not be used except to chase the bond as necessary and in inconspicuous places. The gauge shall be four courses to 300 mm and no brickwork shall rise more than 1000 mm at any one time. The making good of putlog holes and other joint defects, making good etc., shall be in the same mortar mix as that specified for the brickwork as a whole.*
	Block partitions and walls	*Build the ground floor block partitions to the dimensions as shown on the drawings including all openings as shown, all a.b.d. Build in lintels, frames and linings.* *Build the 150 mm thick block wall as shown on the drawings, form openings and build in lintels and frames. Build in all joist hangers and truss supports as required.* *Where brickwork may be subject to frost the written approval of the Architect must be given before any work is carried out.* *Damaged brickwork not so approved or inadequately protected will be re-stated at the Contractor's expense.*